RAND PROJECT AIR FORCE

People's Liberation Army Air Force Operations over Water

Maintaining Relevance in China's Changing Security Environment

Mark R. Cozad, Nathan Beauchamp-Mustafaga

Prepared for the United States Air Force

Approved for public release; distribution unlimited

For more information on this publication, visit www.rand.org/t/RR2057

Library of Congress Cataloging-in-Publication Data is available for this publication.
ISBN: 978-0-8330-9877-1

Published by the RAND Corporation, Santa Monica, Calif.

Preface

This report is based on RAND Project AIR FORCE Strategy and Doctrine Program research that was presented at the second China Aerospace Studies Institute conference, sponsored by Headquarters, U.S. Air Force. It was held at RAND's Arlington, Va., office on May 2, 2016. Experts on airpower, military operations, and Chinese military modernization participated in the conference and provided valuable feedback to this report's authors. The resulting documents assess notable developments in and implications of China's emerging aerospace expeditionary and power-projection capabilities. As China's economic, diplomatic, and security interests continue to expand, the People's Liberation Army (PLA) and in particular its aerospace forces—to include air force, naval aviation, and space capabilities—will require more robust power-projection and expeditionary capabilities on par with China's increasingly global footprint. In addition to traditional security concerns, such as Taiwan and maritime territorial disputes, such issues as countering global terrorism, humanitarian assistance/disaster relief, and sea-lane protection have now become factors in the PLA's training, doctrine, and modernization efforts. In addition, command of space, to include the military use of outer space, is of increasing interest to the PLA as it seeks to develop new capabilities and operating concepts to support its growing range of military missions. This report focuses on PLA Air Force operations over water, a topic with important implications for the United States and its allies and partners in the region.

RAND Project AIR FORCE

RAND Project AIR FORCE (PAF), a division of the RAND Corporation, is the U.S. Air Force's federally funded research and development center for studies and analyses. PAF provides the Air Force with independent analyses of policy alternatives affecting the development, employment, combat readiness, and support of current and future air, space, and cyber forces. Research is conducted in four programs: Force Modernization and Employment; Manpower, Personnel, and Training; Resource Management; and Strategy and Doctrine. The research reported here was prepared under contract FA7014-16-D-1000.

Additional information about PAF is available on our website: www.rand.org/paf

This report documents work originally shared with the U.S. Air Force in May 2016. The draft report, issued on March 1, 2017, was reviewed by formal peer reviewers and U.S. Air Force subject-matter experts.

Contents

Preface ... iii
Summary ... vii
Acknowledgments ... ix
Abbreviations ... xi

1. Introduction ... 1

2. The PLAAF's Expanding Strategic Roles ... 5
   - Beyond a Legacy Framework ... 6
   - Transitioning to a Maritime Focus ... 8
   - Leadership Guidance ... 9

3. Development of a Training Program for Operations over Water ... 13
   - More Realistic Training ... 14
   - The Four Major Brands ... 15
   - New Maritime Combat Training for Cadets ... 18
   - Lessons Learned from Early Experimentation ... 19

4. Recent Training for Operations over Water ... 21
   - Four Flights Past the First Island Chain into the Pacific Ocean in 2015 ... 22
     - Flight Details ... 22
     - Strategic Messaging ... 28
   - "Combat Air Patrols" over the South China Sea in 2016 ... 30
     - Flight Details ... 31
     - Strategic Messaging ... 33
   - Flights into the Pacific Ocean in 2016 ... 36
     - Flight Details ... 36
     - Strategic Messaging ... 38
   - Flights Circumnavigating Taiwan in 2016 ... 39
     - Flight Details ... 39
     - Strategic Messaging ... 41
   - Maritime Strike Training ... 42
   - Potential PLAAF Long-Distance Strike Targets ... 45

5. Joint Service Maritime Training ... 49
   - Sharp Sword-2015 ... 49
   - PLAAF Bomber Coordination with PLAN Ships ... 51

Other Domestic Joint Training 52
Joint Sea-2015 II........ 53
PLAAF-PLAN Air Force Maritime Strike Mission Overlap 54

6. Conclusion 55

References........ 57

Summary

As China's national interests grow globally and the Chinese government seeks to enforce its territorial claims in Asia, the People's Liberation Army (PLA) is increasingly called upon to secure and protect these interests. Overwhelmingly, these interests either reside in or rely on the maritime domain. The PLA Air Force (PLAAF) since its inception has focused its efforts on territorial defense with only limited concern for issues beyond China's mainland. Beginning in 2014, Chinese President and Commander-in-Chief Xi Jinping has led calls for the PLAAF to support PLA efforts to defend China's maritime interests and strengthen its over-water capabilities toward this goal. These efforts are part of a broader effort to prepare for military struggle, particularly in the maritime domain, in part by shaping the security environment to win without fighting. The PLAAF's current modernization initiatives supporting this move include developing long-distance maritime power projection, improving strategic conventional deterrence, and building maritime strike capabilities.

The PLAAF initiated an experimentation and training program in 2014 to meet the new challenges it faces in the maritime domain. This early effort is focused predominately on maritime strike. It includes concept development aided by a combination of increased realism and competitive training for maritime strike missions against regional targets. These new requirements and the PLAAF's responses to the challenges they present provide a lens into the strategic orientation of its future modernization efforts.

Recent PLAAF over-water exercises attempted to tackle these new and challenging problems as demonstrated by four groundbreaking flights into the Pacific Ocean through the First Island Chain in 2015 and flights into the South China Sea and around Taiwan in 2016. By the authors' count, since March 2015, the PLAAF has conducted eight flights past the First Island Chain, including three patrols of the East China Sea Air Defense Identification Zone (ADIZ), two flights around Taiwan, and five flights into the South China Sea. In March 2015, the PLAAF entered the Western Pacific for the first time when H-6K long-range strategic bombers overflew the Bashi Channel, between Taiwan and the Philippines. The Air Force touted this flight as marking an important "breakthrough" of the First Island Chain for projecting power farther into the Pacific. In May 2015, another H-6K unit for the first time overflew the Miyako Strait, near Okinawa, Japan, establishing a second route into the Pacific. In July 2016, the PLAAF began "air combat patrols" in the South China Sea, prominently featuring H-6Ks. Finally, H-6Ks and other support aircraft circumnavigated Taiwan in November and December 2016.

These flights have both operational value for training pilots and political value for strategic messaging. While these flights have been described as normal operations for littoral states and part of the natural development of the Chinese military in official PLAAF announcements, authoritative military commentary suggests that the utility of these flights extends beyond simply training for maritime missions, as they are sometimes also intended to convey strategic signals to relevant countries during times of political tension with China. There is also evidence that this training is laying the groundwork for strikes against regional targets, notably Guam.

The PLAAF has also stepped up joint training with the PLA Navy (PLAN) Air Force since 2015, as its ability to coordinate with the PLAN is one of the central challenges in the PLAAF's new emphasis on maritime operations. This cooperation is most evident in the "Sharp Sword-2015" and "Joint Sea-2015" (with Russia) exercises. Sharp Sword—based on its earlier use as a capstone experimentation and demonstration exercise—presages a likely burst of PLAAF-PLAN joint operations work in the coming years as they share responsibility for safeguarding Chinese interests away from the mainland.

These operations mark a training progression toward increasingly frequent and complex flights and suggest that the PLAAF is transitioning from the experimental phase to regularizing these long-range power-projection activities as PLAAF capabilities mature. Reflecting this, the PLAAF announced in July 2016 that it had "regularized" flights into the South China Sea, and in September 2016 it similarly said that it had "regularized" flights past the First Island Chain. With China's economic and security interests continuing to expand beyond China's immediate periphery, Chinese leaders will likely expect the PLAAF to provide more strategic capacity—enforcing territorial claims, supporting strategic conventional deterrence, and, in the case of war, performing maritime strikes in the region. Although the PLAAF's conceptual development in the maritime arena is in its early stages, PLAAF long-range flight activity in the Pacific will become increasingly common. In addition, as Chinese leaders become more confident in PLAAF capabilities, the service's support to other strategically important initiatives will increase accordingly.

Acknowledgments

The authors would like to thank the directors of the China Aerospace Studies Institute, David Ochmanek and Michael Chase, as well as Lyle Morris, for their guidance and support. The authors would also like to thank their U.S. Air Force counterpart and other reviewers for their constructive comments.

Abbreviations

ADIZ	Air Defense Identification Zone
ASCM	anti-ship cruise missile
CMC	Central Military Commission
DoD	Department of Defense
EW	electronic warfare
GSD	General Staff Department
km	kilometer
LACM	land-attack cruise missile
MND	Ministry of National Defense
MOD	Ministry of Defense
MR	Military Region
PCA	Permanent Court of Arbitration
PLA	People's Liberation Army
PLAAF	People's Liberation Army Air Force
PLAN	People's Liberation Army Navy
PRC	People's Republic of China
TC	Theater Command

1. Introduction

As China's national interests grow globally and the Chinese government seeks to enforce its territorial claims in Asia, the People's Liberation Army (PLA) is increasingly called upon to secure and protect these interests. Overwhelmingly, these interests either reside in or rely on the maritime domain. The PLA Air Force (PLAAF) since its inception has focused its efforts on territorial defense with only limited concern for issues beyond China's mainland. Beginning in 2014, Chinese President and Commander-in-Chief Xi Jinping has led calls for the PLAAF to support PLA efforts to defend China's maritime interests and strengthen its over-water capabilities toward this goal. These efforts are part of a broader effort to prepare for military struggle, particularly in the maritime domain, in part by shaping the security environment to win without fighting. The PLAAF's current modernization initiatives supporting this move include developing long-distance maritime power projection, improving strategic conventional deterrence, and building maritime strike capabilities.

Given the growing importance of the PLAAF's power-projection capabilities, this report provides a comprehensive review of recent known PLAAF training for long-range over-water missions. This research analyzes the operational details and strategic messaging of these flights and places them in the larger context of the PLAAF's strategic transformation toward a maritime focus. This report contributes to the unclassified literature by providing insight into PLAAF training programs and potential long-term strategic goals, including strike targets in the region. Our analysis covers only the PLAAF and not the PLA Navy (PLAN) Air Force, even though its H-6 planes—only the PLAAF has the H-6K—have flown somewhat similar routes through the First Island Chain in recent years. The information cutoff for this report is December 31, 2016.

The remainder of this report has five chapters. Chapter 2 provides an overview of the PLAAF's expanding strategic roles and leadership guidance for a maritime focus. Chapter 3 traces the development of the PLAAF training program for over-water missions, including specific training initiatives to develop specific operational skills. Chapter 4 details recent PLAAF activities to train and operationalize those identified skills, including the strategic messaging from official PLAAF statements and authoritative military commentary. Chapter 5 details PLAAF efforts to deepen interoperability with the PLAN Air Force through joint exercises, and Chapter 6 presents our conclusions.

Our analysis is based on open-source information, and we acknowledge the inherent limitations of this research methodology—we know only what is reported by governments and the media and cannot assess the rigor or realism of PLAAF training, nor could we at times clarify whether each reported flight was unique. While we began our research with PLA and state-run media Chinese-language material, mostly *Kongjun Bao* (officially translated into English as *Air Force News* but better known by its Chinese name), we sought to verify as much reporting as possible. This has been done mostly through corroboration with other foreign government reporting, specifically Japanese Ministry of Defense (MOD) reports and Taiwan Ministry of National Defense (MND) statements. The Japanese MOD reports were the most detailed, as they identified flight paths and provided photographs of each airplane involved, although their coverage was generally limited to the Miyako Strait, meaning that they may not be entirely accurate for longer flights that transit Japanese airspace and continue elsewhere. Furthermore, their reports do not differentiate between PLAAF and PLAN Air Force H-6s, potentially creating some confusion. However, there is no comparable third-party monitoring of the Bashi Channel, as the Philippine and Taiwan governments do not appear to report regularly on Chinese flights through the Channel.[1]

This report focuses mainly on the activities of the PLAAF's newest strategic bomber, the H-6K. Chinese media describe the H-6K as "China's homegrown medium and long range bomber" and note that "it is mainly used for long range precision strikes and tactical bombing missions," with the expectation that it will "play a significant role in safeguarding China both at sea and in the air."[2] The H-6 platform is based on a Soviet bomber design from the 1950s, and the H-6K variant entered service in 2011 and uses new Russian engines and advanced technology to bring the airframe into the modern era. There were reportedly between 13 and 36 H-6Ks deployed in 2016.[3] According to *IHS Jane's 360*, the H-6K is subsonic and has an operational range of roughly 3,000 kilometers (km).[4] It can carry six missiles, including the

---

[1] Although PLAAF aircraft have flown near features in the South China Sea claimed by the Philippines, the Philippines government has not released much information on PLAAF activities in the region. The Taiwan government did not release any detailed information on PLAAF flights near Taiwan in 2015 and only belatedly acknowledged the first PLAAF flight around Taiwan in November 2016. See Frances Mangosing, "PH Can't Confirm Chinese Bomber Flight over Scarborough," *Inquirer*, July 19, 2016; and Lucas Tomlinson and Jennifer Griffin, "China Flew Nuclear-Capable Bombers Around Taiwan Before Trump Call with Taiwanese President," Fox News, December 5, 2016.

[2] For a representative sample of Chinese media reporting on the H-6K, see Zhang Tao, "Expert: China's Bombers Embrace New Opportunities," China Military Online, February 25, 2015; Xu Sheng [徐生], Zhao Lingyu [赵凌宇], and Wang Wenbin [王文彬], "Shocking the Skies by Piloting the 'Gods of War'" ["驾'战神'以撼苍穹"], *PLA Daily*, December 8, 2015, p. 5; Yao Jianing, "H-6K Bomber, Ace of PLA Air Force," People's Daily Online, June 6, 2016; Yao Jianing, "Expert: PLA Air Force Can Cover Whole South China Sea," China Military Online, July 19, 2016; and Zhang Tao, "PLA Air Force's South China Sea Patrol a 'New Normal,'" China Military Online, August 25, 2016.

[3] National Institute for Defense Studies, *NIDS China Security Report 2016: The Expanding Scope of PLA Activities and the PLA Strategy*, March 2016, p. 27; and "China—Air Force," *Jane's World Air Forces*, December 23, 2016.

[4] Other sources put the H-6K's range at 3,500 km. See Jordan Wilson, "China's Expanding Ability to Conduct Conventional Missile Strikes on Guam," U.S.-China Economic and Security Review Commission, May 10, 2016; "XAC H-6," *IHS Jane's 360*, undated; and Richard Fisher, "China Showcases New Weapon Systems at 3 September Parade," *Jane's Defence Weekly*, September 8, 2015.

Chinese media often describe a range of 8,000 km and 10 hours of flying time, which is similar to Western estimates, assuming Chinese estimates are for ferry range. See Yao Jianing, July 19, 2016.

CJ-20 land-attack cruise missile (LACM) with a range of 1,500 km, giving it a 4,500-km strike range. Although the H-6K is often referred to as nuclear-capable, "China's H-6 bomber is not nuclear-capable" at this time, according to Jeffrey Lewis.[5] The Department of Defense (DoD) 2016 report on the Chinese military adds, "China might eventually develop a nuclear bomber capability."[6] This is despite one older H-6 variant being used to conduct several nuclear tests in the early years of China's nuclear program and a 2013 U.S. Air Force report that described the CJ-20 as nuclear-capable.[7]

---

[5] Lucas Tomlinson, "China Flies Nuclear-Capable Bomber in South China Sea After Trump Taiwan Call, US Officials Say," Fox News, December 9, 2016; Erik Slavin, "Does China Really Have a Nuclear-Capable Bomber?" *Stars and Stripes*, December 13, 2016; and Kyle Mizokami, "China's Flyover Message to Trump Wasn't Quite What It Seemed," *Popular Mechanics*, December 12, 2016.

[6] Office of the Secretary of Defense, *Annual Report to Congress: Military and Security Developments Involving the People's Republic of China 2016,* Arlington, Va.: U.S. Department of Defense, April 26, 2016, p. 38.

[7] Slavin, December 13, 2016; and James M. Kowalski, "Air Force Global Strike Command," U.S. Air Force, May 7, 2013.

2. The PLAAF's Expanding Strategic Roles

China's rise brings with it numerous strategic imperatives and concerns. These include expanded economic engagement in diverse regions, growing political influence and responsibilities, and new challenges to its territorial claims. China's leaders are calling on the military to secure and protect these interests. PLA modernization efforts thus are driven by a renewed sense of urgency and senior-level interest. In a November 2015 speech, PLAAF Commander Ma Xiaotian argued that "because our national development, maritime rights protection, foreign economic activities, and non-war military actions are increasing by the day, it will be necessary for us to further increase our awareness of the urgency of making preparations for maritime military conflict properly."[8]

These imperatives also are driving the PLA's services to develop capabilities directed toward new missions in unpredictable and uncertain environments.[9] Future conflicts require flexible and rapidly deployable capabilities ready to address contingencies in multiple domains. The People's Republic of China's (PRC's) expanding global engagement eventually will force all PLA services to adapt the way they plan, deploy, and fight. To differing extents, all PLA services currently are dealing with these new requirements. Senior-level directives have dictated the methods under development and on display in exercises and other training venues. Furthermore, the PLA's recent reorganization is aimed at ensuring a more nimble, operationally oriented force by stripping away the PLA's legacy structures oriented toward territorial defense of China's mainland. Chinese leaders intend for these organizational and operational reforms to enable the PLA to "effectively control major crises, properly handle possible chain reactions, and firmly safeguard the country's territorial sovereignty, integrity and security."[10]

[8] Yang Qingchun [杨庆春], "While Inspecting a Certain Guangzhou Military Region Air Force Unit, Ma Xiaotian, Member of the Central Military Commission and Commander of the Air Force, Emphasized That Joint Maritime Search and Rescue Will Require Integration of Military and Local Organizations and High Command Efficiency" ["中央军委委员, 空军司令员马晓天在广空部队检查调研时强调:海上联合搜救要军地-体多元融合指挥高效"], *Kongjun Bao*, November 10, 2015.

[9] Information Office of the State Council, *China's Military Strategy*, reprinted by Xinhua [新华], May 26, 2015, p. 11. In addition, there have been numerous exercises dating back to at least 2005 that have addressed these new missions and the environmental conditions in which they are undertaken. Key exercise series include Sharp Sword, Mission Action, Strike, and Firepower—all of which have been performed over multiple years, with a capstone demonstration performed at the end of its designated Five Year Plan. See Mark R. Cozad, "PLA Joint Training and Implications for Future Expeditionary Capabilities: Testimony Presented Before the U.S.-China Economic and Security Review Commission, January 21, 2016," Santa Monica, Calif.: RAND Corporation, CT-451, 2016.

[10] Information Office of the State Council, 2015.

Beyond a Legacy Framework

The PLAAF since its inception has focused its efforts on territorial defense with only limited concern for issues beyond China's mainland. Over the past two decades, there has been a growing recognition among China's military and civilian leaders that the PLAAF's capabilities need to be strategically oriented—a major departure from its historically territorial defense-minded approach. China's most recent military strategy, contained in its 2015 Defense White Paper, highlighted the degree to which the PLAAF is looking beyond its traditional orientation.[11] Over the past 15 years, the PLAAF has spent considerable time developing its "strategic air force" concept.[12] The ideas that emerged from this process have been a major influence on recent PLAAF thinking, shaping its views on primary missions and operational concepts. According to the National Air and Space Intelligence Center, "analysis of PLAAF writings [since 2002] indicate that the PLAAF's air tactics training have ranged from simple formation flying to nighttime, long-range, and minimum-altitude attacks over water against maritime and ground targets."[13]

Beginning in 2004, the Central Military Commission (CMC) put forth the strategic requirement for the PLAAF to perform air-space integration along with both attack and defense capability.[14] In the same directives, the CMC expanded the PLAAF's strategic missions by giving it the additional missions of supporting PLA Army and PLAN operations.[15] An authoritative 2009 PLAAF publication, based on the makeup of its editorial committee and its wide usage in military circles, contained added detail on what a strategic air force should be capable of performing: "the development of national interests overseas calls for the Air Force to gradually possess certain capabilities to perform overseas operations."[16] This included long-range reconnaissance and surveillance, supporting joint maritime blockade, conducting long-range air transport and mobility, and carrying out long-range precision strikes outside China's borders. China's most recent military strategy provides further details in the context of PLAAF operational requirements to meet the needs of informatized warfare by emphasizing capabilities

---

[11] "In line with the strategic requirement of building air-space capabilities and conducting offensive and defensive operations, the strategy PLA Air Force (PLAAF) will endeavor to shift its focus from territorial air defense to both defense and offense, and build an air-space defense force structure that can meet the requirements of informatized operations." See Information Office of the State Council, May 26, 2015, p. 6.

[12] For a more in-depth discussion of China's strategic air force concept, see Michael S. Chase and Cristina L. Garafola, "China's Search for a 'Strategic Air Force,'" *Journal of Strategic Studies*, Vol. 39, No. 1, September 14, 2015, pp. 4–28.

[13] National Air and Space Intelligence Center, *People's Liberation Army Air Force 2010,* Dayton, Ohio, 2010, p. 82.

[14] Shou Xiaosong [寿晓松], ed., *The Science of Military Strategy* [战略学], Beijing, China: Military Science Press [军事科学出版社], 2013, p. 220.

[15] Shou Xiaosong, 2013.

[16] Zhu Hui [朱晖], ed., *Strategic Air Force* [战略空军论], Beijing, China: Lantian Publishing House [蓝天出版社], 2009, p. 46.

in several core areas, including "strategic early warning, air strikes, air and missile defense, information countermeasures, airborne operations, strategic projection and comprehensive support."[17] Last, the 2016 DoD report on the Chinese military asserted that the PLAAF was given a deterrence mission in 2012, and "some Chinese military analysts have described [long-range bombers] as 'capable of performing strategic deterrence.'"[18]

Indeed, PLAAF training in 2015 and 2016 included all of these missions from the 2015 military strategy except explicit training for a maritime blockade. The importance of the maritime domain for all these missions is evident in one Chinese review of PLAAF activities in 2015 that hailed "seven breakthroughs," of which four had over-water components.[19] These breakthroughs were increased distant sea operational capability, large-scale training, the ability to win future naval warfare, and aerial refueling throughout Chinese territory. This training is detailed in Chapter 4.

In practice, the PLAAF's mandate to develop a strategic air force has been tied overwhelmingly to missions over land. Building acceptance within the PLAAF to pursue missions over water thus has proven difficult.[20] As one observer noted, these integration problems are not specific to the PLAAF. He noted that generally "relevant countries and relevant domestic departments are 'not used to' the Air Force departing the border, and categorically see it as a threat and provocation."[21] Within the PLAAF, this deeply ingrained perspective limited the PLAAF's ability and willingness to participate in maritime training; however, this reluctance and resistance was unsustainable in light of the PRC's new strategic situation.[22]

President Xi broke this barrier forcefully in 2014 when he described the PLAAF as a strategic service, thereby expanding its reach to support China's maritime interests.[23] The PLAAF subsequently initiated an experimentation and training program to meet this challenge.

---

[17] Information Office of the State Council, May 26, 2015, p. 6.

[18] Office of the Secretary of Defense, April 26, 2016, p. 38.

[19] The other three breakthroughs were more advanced early warning aircraft, an improved anti-aircraft fires system (namely, surface-to-air missiles), and the inclusion of female pilots in the "main battlefield." A similar review of 2016 training listed five H-6K flights. See Huang Zijuan [黄子娟], "Chinese Air Force's Combat Ability Had Seven Breakthroughs: Aimed at Winning Future Maritime Warfare" ["中国空军近年七大突破:远海作战能力大幅提升"], People's Daily Online, November 11, 2015; and He Wu [鹤舞], "Wow! The Air Force Did All This in One Year Since Military Reform" ["厉害!军改一年空军干了这么多大事"], China Military Online, November 25, 2016.

[20] PLAAF efforts to improve over-water capabilities date back to at least 2004, when the *PLA Daily* hailed "breakthroughs made in maritime night time flying." This was considered an "[indication of] the willingness to accept the likelihood of significantly increased operational training losses in order to begin building an experience base among PLAAF fighter and attack aircraft crews." See Roger Cliff, John F. Fei, Jeff Hagen, Elizabeth Hague, Eric Heginbotham, and John Stillion, *Shaking the Heavens and Splitting the Earth: Chinese Air Force Employment Concepts in the 21st Century,* Santa Monica, Calif.: RAND Corporation, MG-915/2-AF, 2011, p. 217.

[21] Zhu Hui, 2009, p. 35.

[22] Zhu Hui, 2009, p. 35.

[23] "Xi Jinping Inspects Air Force Organ" ["习近平在空军机关调研"], April 14, 2014, Xinhua; Chase and Garafola, 2015, p. 1.

This important first step, which remains in progress, includes concept development supported by a combination of increased realism and competitive training for maritime strike against regional targets. These new requirements and the PLAAF's responses to these challenges provide a lens into the future direction of its modernization efforts.

Transitioning to a Maritime Focus

Overwhelmingly, China's leaders recognize that the nation's interests—along with the major threats to them—reside in the maritime domain. President Xi's efforts to mitigate and confront these threats underpin his recent directives for the PLA to prepare for military struggle.[24] These preparations proceed from a recognition that modern forms of warfare and the current international security situation require that the PLA improve its capabilities in this critical domain. China's most recent military strategy points out that the ability to fight and win informatized local wars requires advanced capabilities in the maritime environment.[25] The PLAAF's preparations for future military struggles are tied closely to this emerging operational need. They likely will remain so for the foreseeable future.

Preparing for maritime military struggle is a long-term strategic task composed of conceptual development, experimentation, and force-wide training, according to Senior Colonel An Peng, head of research at the PLA Air Force Command College in Beijing.[26] Writing in the PLA's most prominent journal, *China Military Science*, he asserts that the PLAAF's efforts toward preparing for maritime military struggle build on previous PLAAF efforts to develop a strategic air force capable of supporting China's national security objectives beyond mainland defense. These efforts to develop a strategically relevant air force largely arose from a recognition that future wars would rely increasingly on air power to achieve their objectives. According to an

---

[24] The PLA defines *military struggle* as "using mainly military means to engage in combat with countries or political groups for certain political, economic or other goals. The highest form is war," and defines *preparing for military struggle* as "engaging in preparations for fulfilling the requirements of military struggle. The core is preparations for war." One analyst said, "preparing for military struggle" is "similar to the concept of operational readiness." See M. Taylor Fravel, "No, Hu Didn't Call for War," *The Diplomat*, December 10, 2011; and *PLA Military Terms* [军语], Beijing, China: Military Science Publishing House [军事科学出版社], December 2011, p. 5.

[25] Although this is the first document labeled and published as a military strategy, it follows a biannual series of Defense White Papers that have outlined China's threats, its military structure, and the military's role in supporting these national security objectives. For analysis on the report, see M. Taylor Fravel, "China's New Military Strategy: 'Winning Informationized Local Wars,'" *China Brief*, July 2, 2015.

[26] An Peng [安鹏], "Strategic Consideration on Strengthening the Air Forces in the Maritime Direction" ["加强海上方向空中力量建设的战略思考"], *China Military Science* [中国军事科学], No. 3, 2015, pp. 82–85.

An Peng has also played a leading role in other forward-leaning research for the PLAAF, including on joint operations and "system-of-system" operations. See Ma Dezu [马德祖], Yang Chunyuan [杨春源], and Zhao Feipeng [赵飞鹏], "Practice to Command a War: Report on the Joint Exercises Conducted by the PLA Air Force Command Academy's New Graduates" ["练习指挥一场战争: 空军指挥学院毕业学员联合作业演习实录"], China Youth Daily Online, January 18, 2008; "Military Report," China Central Television, November 2, 2010; and "Military Report," China Central Television, August 1, 2014.

authoritative PLA book, modern wars have demonstrated that seizing command of the seas is one of three essential elements to success, along with information and air superiority—together known as the "three superiorities," a central concept in PLA operational thinking for a variety of campaigns and operational scenarios.[27] This observation is particularly relevant to China's military planners focused on Taiwan, the East China Sea, or the South China Sea. More important, according to An, the "experience of modern war shows that the key to seizing command of the seas is seizing command of the air."[28]

Two major strategic motives emerge from this realization—both of which link the PLAAF's over-water capabilities development to China's broader strategic interests. First is the recognition that maritime powers require capable air power for use in the maritime domain. Second is that effective integration of air and naval forces is a key determinant for victory in naval warfare.[29] The PLAAF's new emphasis on building capabilities for supporting warfare in the maritime domain provides the second of two strategic orientations guiding future capabilities developments: one land oriented and the second focused on the "sea border" and the maritime environment. Subsequent modernization and training efforts are, in turn, aimed at developing a strategic air force that can respond rapidly, deploy at long ranges, and operate in both offensive and defensive capacities. If the PLAAF is successful in building these capabilities, it will be able to perform effectively its missions in both "strategic directions" (i.e., the land border and the sea border) and allow the PRC to protect its contested maritime territory.[30]

Leadership Guidance

High-profile guidance and support from President Xi, and from other senior military leaders, reflect the PLA's sense of urgency for this expanding mission set. President Xi's April 2014 visit to PLAAF headquarters in Beijing for the first time embraced the Air Force as a strategic service, and his February 2015 visit to a PLAAF bomber unit near Xi'an highlighted the role of the H-6K, the PLAAF's newest and most capable long-range strategic bomber.[31] CMC Vice

---

[27] Zhang Yuliang [张玉良], ed., *The Science of Campaigns* [战役学], Beijing, China: National Defense University Press [国防大学出版社], 2006, p. 80.

[28] An Peng, 2015, pp. 82–85.

[29] An Peng, 2015, pp. 82–85.

[30] An Peng, 2015, pp. 82–85.

[31] President Xi also visited a PLAAF base in February 2013, but the emphasis was on weapon development and did not focus on maritime operations. For details on how the PLAAF interpreted Xi's visit for training guidance, see Yu Guozhong [于国忠] and Xu Tongxuan [徐同宣], "The PLA Air Force Convenes Its Semi-Annual Video Teleconference on Military Training; Ma Xiaotian, Member of the Central Military Commission and Air Force Commander, and Air Force Political Commissar Tian Xiusi Are Present" ["空军召开半年军事训练电视电话会议：中央军委委员，空军司令员马晓天，空军政治委员天修思出席"], *Kongjun Bao*, June 1, 2014; Wang Shibin, "Xi Jinping Visits Officers and Men of PLA Air Base," China Military Online, February 5, 2013; Chase and

Chairman General Fan Changlong's October 2014 visit to an unidentified PLAAF command post included discussion of "training for . . . distant sea" activities, and his July 2016 visit to a PLAAF base in the Southern Theater Command (TC), which appears to combine parts of the former Guangzhou and Chengdu Military Regions (MRs), emphasized the need to prepare for military struggle in the maritime direction.[32]

PLAAF Commander Ma Xiaotian has repeatedly stated the importance of improving the PLAAF's over-water capabilities, including in an April 2014 speech outlining party guidance for strengthening capabilities to deal with fighting and winning wars and discussing the safeguarding of maritime rights and interests.[33] Ma pointed out that PLAAF operations to safeguard PRC maritime rights and interests brought new and complex requirements in multiple areas. One component was focused on a conceptual perspective of making PLAAF commanders and operators "fully and clearly aware of the importance of winning the initiative in air struggle" and embedding a sense of urgency for "effectively coping with various security threats from the maritime domain." Training to meet these new missions called for a sense of urgency for the PLAAF to quicken the pace of transforming itself from its primary focus on homeland air defense to possessing both offensive and defensive capabilities that could be employed well beyond China's land borders and into the maritime domain. Ma further stated that this had been a weak area for the PLAAF traditionally, and moving away from its legacy perspectives would require additional top-down direction. He told senior PLAAF leaders that they must

> Continuously improve the air power system in the maritime domain by focusing on the problems. We should tightly fix our eyes on the weaknesses in our air

---

Garafola, 2015; Richard Fisher, "Xi Jinping Visit Reveals H-6 Bomber Details," *IHS Jane's Defence Weekly*, February 19, 2015; and Zhang Tao, February 25, 2015.

[32] In February 2016, the PLA reorganized its military from seven MRs to five TCs, acting on reforms announced in late 2015 and under discussion within the PLA for many years. This report will use the appropriate designation depending on the time referenced.

For a good discussion regarding what is known about the reforms, see Kenneth W. Allen, Dennis J. Blasko, and John F. Corbett, Jr., "The PLA's New Organizational Structure: What Is Known, Unknown and Speculation (Part 1)," *China Brief*, February 4, 2016; Kenneth W. Allen, Dennis J. Blasko, and John F. Corbett, Jr., "The PLA's New Organizational Structure: What Is Known, Unknown and Speculation (Part 2)" *China Brief*, February 23, 2016; Xiong Weiwei [熊伟威], "Investigating and Researching at Air Force Organs, Fan Changlong, Member of the CPC Central Committee Political Bureau and Vice Chairman of the Central Military Commission, Stresses the Serious Implementation of the Series of Important Decisions and Instructions of Chairman Xi and Making Persistent Efforts in Deeply Pushing Forward Actual Combat-Oriented Training: Fang Fenghui, Member of the Central Military Commission and Chief of General Staff, Conducts Research and Study, Ma Xiaotian, Qi Jianguo, Wu Changde, Liu Zheng, Liu Sheng, and Tian Xiusi Also Participate in the Research" [中共中央政治局委员、中央军委副主席范长龙在空军机关调研时强调:认真贯彻习主席系列重要决策指示再接再厉深入推进实战化训练], *Kongjun Bao*, October 27, 2014, p. 1; and "Fan Changlong: Continue Deepening Preparations for Military Struggle in the Maritime Direction, Effectively Carry Out the Historic Missions Assigned by the Party and People" ["范长龙:不断深化细化海上方向军事斗争准备 有效履行党和人民赋予使命任务"], Xinhua, July 19, 2016.

[33] Ma Xiaotian [马晓天], "Strive to Enhance the Air Force's Capability of Fighting and Winning" ["努力提高空军部队能打仗打胜仗能力"], *PLA Daily* [解放军报], April 2, 2014, p. 6.

> actions in the maritime domain, concentrate efforts on making breakthroughs in solving crucial and difficult issues. We should persistently take information as a dominating factor, energetically strengthen the building of information support power; stress system-on-system conflict, strive to enhance the air offensive and defensive system-of-systems operations in distant seas areas; push forward civil-military integration, and continuously improve the basic support for the use of the Air Force power.[34]

Ma's statement tied training for maritime missions closely to other areas that had featured prominently in PLAAF modernization efforts in recent years—informatized warfare, system-of-systems operations, and comprehensive support.[35] As these examples demonstrate, high-level leadership engagement is an essential element of improving the PLAAF's ability to operate in multiple roles over water. In light of this increased senior interest, the PLAAF is faced with developing operational concepts and tactics that will make these directives concrete within the PLAAF's playbook and has endeavored to accelerate the transition toward a maritime focus.

---

[34] The system-of-systems concept "has its roots in U.S. military writings on network-centric warfare and involves combat between systems of systems rather than between individual systems or platforms." For more details, see Kevin L. Pollpeter, Michael S. Chase, and Eric Heginbotham, *The Creation of the PLA Strategic Support Force and Its Implications for Chinese Military Space Operations,* Santa Monica, Calif.: RAND Corporation, RR-2058-AF, 2017; Kevin McCauley, "System of Systems Operational Capability: Key Supporting Concepts for Future Joint Operations," *China Brief,* October 5, 2012; Kevin McCauley, *PLA System of Systems Operations: Enabling Joint Operations,* Washington, D.C.: Jamestown Foundation, 2016; and Ma Xiaotian, April 2, 2014, p. 6.

[35] Zhang Yuliang, 2006, p. 80.

3. Development of a Training Program for Operations over Water

The PLAAF has developed a coherent training program to support its growing responsibilities in the maritime domain in response to calls by Chinese leaders. This includes an increase in the overall realism of combat training, a framework known as the "Four Major Brands" for training core skills required to successfully conduct PLAAF missions, and even new textbooks for cadets at its flight school. These efforts resulted in a September 2015 meeting to review early progress in the training program.

President Xi's directives were focused on two essential areas. The first was ensuring that the PLA as a whole and the PLAAF in particular understood the growing importance of the maritime domain to China's strategic interests. Second, the PLAAF needed to address new challenges within this domain that had traditionally been dominated by the United States—the main enemy discussed in PLA military science literature. Accordingly, the PLAAF's importance for ensuring China's military capability in the maritime domain became a central theme in the PLAAF's preparation for military struggle.

PLA researchers have stated in the most fundamental of terms "whoever can seize air superiority can then gain greater strategic initiative in the maritime military struggle."[36] Airpower's primary mission supporting operations in the maritime domain are thus centered on seizing and maintaining superiority in the air battle. Although this realization is not a major departure from previous PLA discussions regarding the "three superiorities," it does represent an element of thinking about power projection that had not previously been at the forefront of PLA strategy. Airpower may have been viewed as a key tool for projecting force in the past, but it generally lacked a specific orientation. President Xi's focus on the maritime domain added an essential piece of direction for PLAAF training beyond missions directed against Taiwan.

The PLAAF's new emphasis on preparing for military struggle in the maritime domain does not relieve it from its more traditional areas of focus, most notably Taiwan and homeland air defense. The PLAAF is still expected to focus on the "main strategic direction"—i.e., Taiwan—by maintaining its readiness to conduct both offensive and defensive support to core operational missions, including warning/demonstrative fire strikes, joint fire strikes, blockade operations, island-landing operations, and theater air defense.[37] These preparations are supposed to be done jointly and in coordination with the other PLA services and arms. The PLAAF's second core mission area for preparing for military struggle includes organizing and conducting peacetime

[36] An Peng, 2015.

[37] Shou Xiaosong, 2013, p. 221.

and wartime homeland air defense operations.[38] Both of these core areas have been a central component of PLAAF training and experimentation for several years.

Training for missions in the maritime domain presents the newest challenge to PLAAF readiness. This third component for preparing for military struggle includes developing capabilities to "safeguard security, stability and maritime rights and interests in border- and coastal-defense areas" through the defense of China's coastal areas and protecting PRC maritime rights and interests.[39]

More Realistic Training

PLA training reform based on President Xi's guidance to prepare for military struggle has begun to move at multiple levels and includes the issuance of new training guidance. Most notably, at the beginning of 2014, the CMC released the *Opinions on Raising the Level of the Realistic Battle Orientation of Training,* and at the beginning of 2015, the General Staff Department (GSD) specially issued the *Opinions on Strengthening and Improving Campaign and Tactical Training*.[40] These new guidelines reinforced several key training themes that Chinese leaders had previously put forth, including the need for developing long-range power projection and the capability to "destroy the enemy's space-based information support systems, command information systems, large naval platforms, naval and air bases, and other vital points and weaken the functionality and warfighting potential of the enemy's battle systems."[41] In addition, the development of these capabilities needed to be directed toward gaining some level of balance against the United States, deterring other regional powers, and protecting key targets in China. Last, these guidelines built on PLAAF Commander Ma's guidance to "energetically strengthen combat-realistic training in the maritime domain by closely linking training with tasks."[42] From a systemic perspective, the PLAAF's new direction in training was clear—a strategic air force also needed to be trained in the maritime domain.[43] The progression of such high-level directives on increasing combat-realistic training may indicate either that the CMC was not satisfied with the current training programs or that there was resistance to such reforms.

---

[38] Shou Xiaosong, 2013, p. 221.

[39] Shou Xiaosong, 2013, p. 221.

[40] Xu Tongxuan [徐同宣], "Where Does Training Go from Here? Viewing Realistic Battle Training During Informatized Transformation from the Perspective of Air Force Defense Penetration and Assault Competitive Assessments" ["训练向哪里去:从空军突防突击竞赛性考核看信息化转型时期实战化训练"], *Kongjun Bao*, July 16, 2015, p. 3.

[41] An Peng, 2015, pp. 82–85.

[42] Ma Xiaotian, April 2, 2014, p. 6.

[43] For specific guidance on PLAAF training in 2016, see Xu Tongxuan [徐同宣], "It Is Particularly Essential to Step Up Military Training and Combat Readiness in Times of Reform and Tackling Tough Issues: Sidelights on the PLA Air Force's 2016 Military Training Conference" ["越是改革攻坚越要练兵备战:空军 2016 年度军事训练会议侧记"], *Kongjun Bao*, January 6, 2016, p. 1.

The Four Major Brands

One prominent concept the PLAAF has developed to encapsulate its recent training focus is the Four Major Brands (四大品牌). While there have been various formulations of the concept as it evolved following then-PLAAF Political Commissar Tian Xiusi first hinting at it in a June 2014 speech, the concept incorporates training for four components, namely system-of-systems confrontations, penetration and assault (strike), air combat, and the missile forces.[44] The concept appears to have been developed in response to several recent high-level leadership calls to action discussed above, including President Xi's April 2014 speech on the PLAAF's development, the aforementioned training directives from the CMC and GSD in 2014 and 2015, China's most recent military strategy, and the CMC's 2004 requirement for the PLAAF to integrate air and space as well as attack and defense capabilities.[45]

The PLAAF finalized the Four Major Brands concept in mid-2016 to now include regular training exercises that correspond to each component, specifically the "Red Sword," "Golden Dart," "Golden Helmet," and "Blue Shield" exercises.[46] The evolution of the concept likely reflects the service's gradual refinement of early efforts to respond quickly to leadership demands for improved training. The system-of-systems confrontation training is now embodied by the Red Sword exercises, which have been held since 2005 and are "systematic confrontation [exercises]" that "feature multi-day, large-scale events that include all branches of the PLAAF in opposing force scenarios. These events are undoubtedly improving PLAAF combined arms capabilities and better preparing them for combat."[47] Golden Dart defense penetration and air-to-

---

[44] Tian Xiusi [田修思], "Strive to Quicken the Building of a Strong People's Air Force with Integration of Air and Space and with Both Offensive and Defensive Capabilities Around the Fulfillment of the Party's Military Strengthening Goal Under the New Situation" ["紧紧围绕实现党在新形势下的强军目标为加快建设空天一体攻防兼备的强大人民空军而奋斗"], *Kongjun Bao*, June 24, 2014, pp. 1–2.

Of note, Political Commissar Tian retired and was subsequently investigated for corruption. See Minnie Chan and Choi Chi-yuk, "Former Top Officer in China's Air Force 'Under Investigation by Anti-Corruption Officials': Sources," *South China Morning Post*, July 8, 2016.

[45] Ka Weixing [卞卫星], "Tian Xiusi Observes Blue Sky Aerial Competition and Awards" ["田修思观看空装篮球决赛并颁奖"], *Kongjun Bao*, June 16, 2015, p. 1.

[46] In February 2016, the PLAAF said that 2016 training for "four major brands" would be "opposing-forces air battle competitive exams; competitions in penetrating defenses to attack; 'Red Sword' exercises; and 'Blue Shield' live-fire tactical exercises." See Li Jianwen [李建文], "The Trends of Military Training in the New Year" ["新年度军事训练趋势扫描"], *PLA Daily*, February 1, 2016; and Yang Weike [杨伟科] and Li Jianwen [李建文], "'Golden Dart' Competition Starts Again" ["'金飞镖'争夺战烽烟再起"], *PLA Daily*, July 23, 2016.

[47] "PLA's First Symposium on Military Flight Training Concluded," China Military Online, October 16, 2014; Lee Fuell, "Broad Trends in Chinese Air Force and Missile Modernization," Testimony Before the U.S.-China Economic and Security Review Commission, January 30, 2014; David Axe, "China's Increasingly Good Mock Air Battles Prep Pilots for Real War," *Wired*, February 7, 2013; Sun Maoqing [孙茂庆], Xu Zhuangzhi [徐壮志], and Li Xuanliang [李宣良], "Vast Desert Sky Quakes with Thunder: Air Force Test and Training Base Forges Modern Air Force" ["大漠长天听惊雷:空军某试验训练基地淬火现代空军纪实"], Xinhua, December 1, 2008; Li Jianwen, 2016; and Xu Bianchun [徐边春] and Yang Chaoli [杨朝钊], "While Part of the Force Returns from Triumphant

ground attack assessment drills began in 2014 and are "mainly organized in aviation units that shoulder ground-directed and sea-directed attacks," after penetration and assault training began by at least 2008 and competitions began in 2010.[48] Golden Helmet air combat exercises began in 2011 and are "mainly organized in the fighter plane units, through freestyle air fight with similar plane models and heterogeneous plane models," and focus on "air control operation capabilities."[49] Blue Shield exercises focus on the PLAAF's missile forces and have taken place annually since at least 2013.[50]

For PLAAF over-water activities, the most relevant component is penetration and assault training. China's 2015 Defense White Paper said, "The Air Force is to improve the capabilities of strategic early warning, air strike, [and] strategic projection," among others, so penetration and assault training fulfills this task.[51] Ma made this apparent during his remarks at the September 2015 maritime penetration and assault conference, saying "penetration and assault contest and evaluation [are] one of the Air Force's 'four brand names' for combat-realistic training," referencing training for all the coastal military regions except Guangzhou.[52] Reflecting efforts for more emphasis on maritime training, the Golden Dart exercise was "organized in the sea direction" for the first time in September 2015 to "[sharpen] units' sea combat capability."[53] Reinforcing the force-wide nature of the maritime missions, the winners from the Golden Dart

---

Air Battle, the Other Part of the Force Sets Off to Maritime Training" ["这边空战凯歌还 那边海训再出征"], *PLA Daily*, December 17, 2015, p. 9.

[48] For more information on Golden Dart, see Michael S. Chase, Kenneth W. Allen, and Benjamin S. Purser III, *Overview of People's Liberation Army Air Force "Elite Pilots,"* Santa Monica, Calif.: RAND Corporation, RR-1416-AF, 2016, pp. 12–13; Tan Jie [谭洁], "Nanjing Air Force Relies on Large-Scale Exercise to Accelerate Training Transformation: Air Division Completely Enters Ranks of 1st Class Unit in Military Training" ["南空依托大型演练加速推进训练转变:所属航空兵师全部跨入军事训练一级单位行列"], *PLA Daily*, April 4, 2008; and Li Qing [李清], "Chinese Air Force Conducts Golden Dart Ground Attack Exercise for the First Time" ["中国空军首次举行对地攻击"金飞镖"大赛"], China Military Online, October 7, 2014.

[49] For more information on Golden Helmet, see Chase, Allen, and Purser, 2016, pp. 9–16; Huang Zijuan, November 11, 2015.

[50] Qiu Yue [邱越] and Zhao Yanhong [赵艳红], "'Blue Shield-2014' Exercise: An Air Force Missile Brigade Sharpens Its Sword in the Gobi Desert" ["'蓝盾—2014'演习: 空军某地空导弹旅戈壁大漠引弓砺剑"], China Military Online, August 14, 2014; and Liu Feng'an [刘逢安], "Space Net Won't Leak" ["天网恢恢疏而不漏"], *Modern Military* [现代军事], November 2013

[51] Information Office of the State Council, May 26, 2015.

[52] Xu Tongxuan [徐同宣], "PLA Air Force Holds a Meeting in Beijing to Review and Reflect upon Performance in the Maritime Penetration and Assault Contest and Evaluation Events, Stressing the Settlement of Crucial and Difficult Issues in Force Transformation and Building, Calling for More Elaborate Training Guidance" ["空军在京召开海上突防突击竞赛考核论反思会强调:解救转型建设重难点问题提高训练指导精细化水平,马晓天于忠福全程参与交流研讨并诩话"], *Kongjun Bao*, September 17, 2015, p. 1.

[53] The 2016 Golden Dart exercises do not appear to have had an over-water component. See Yang Weike and Li Jianwen, 2016; Tao Shelan [陶社兰], "The Chinese Air Force Strengthens Actual Combat-Oriented Air Fights Competition, and One Batch of 'Golden Helmet' Pilots Emerge" ["中国空军加强实战化空战比武 一批"金头盔"飞行员脱颖而出"], Xinhua, September 8, 2015.

exercise were from the Lanzhou MR, which is now mostly the Western TC.[54] While the PLAAF has been conducting over-water strike training for a long time, and doing so over land since its creation, the inclusion of such training as one of the four core training components at a time when PLAAF exercises have increasingly shifted to a maritime focus reflects the PLAAF's overall emphasis on bolstering all capabilities in the maritime domain, especially maritime strike.

In reviewing the past five years of PLAAF modernization efforts in his June 2014 speech, Political Commissar Tian noted progress toward more realistic training and created the central framework for what would later become the Four Major Brands. He called for the PLAAF to conduct the necessary "training required by actual combat," including "system-of-systems confrontation training," and "let exercises, free air dogfight drills, penetration and assault contests and tests shape brand-name forces and play an important driving role."[55] PLAAF Commander Ma echoed this training framework in December 2014 on a visit to a training base by saying that "for the issue of 'training,' at present, it is necessary to enhance aerial dogfight drills, force-on-force exercises, penetration and assault tests based on competitions, and the implementation of such brand-name combat-realistic training activities as the Blue Shield."[56] Commander Ma identified "problems . . . in the training system" so the "fundamental requirement" of training reform is to "boost effectiveness and lower depletion," including "[making] the capability of pilots in the combat units satisfy the operation requirements, [and bringing] the overall training standard into line with the requirements of the Air Force's missions and tasks." As the PLAAF's mission has expanded to include maritime activities, training must now increasingly be conducted over the sea.

A series of *Kongjun Bao* articles in June and July 2015 provided the first detailed explanation of the Four Major Brands concept. It clarified that the concept is intended to "establish the single fundamental standard of combat capabilities . . . for the sake of successfully fighting battles."[57] System-of-systems confrontation exercises make soldiers realize they "must be courageous," as "informatized warfare . . . requires . . . a spirit of utter fearlessness" and that they "must not fear sacrifice." This was especially a concern for younger soldiers, born after 1980, who grew up in

---

[54] Fan Jun [范军] and An Hongxin [安红欣], "Lanzhou Military Region Air Force Brigade's 'Golden Dart' Aviators Seek the Way to Fight and Win" ["向精学胜战突进:兰空航空兵某旅'金飞镖'飞行员谋打赢之道"], China Air Force Online [中国空军网], January 26, 2016.

[55] Tian Xiusi, June 24, 2014, pp. 1–2.

[56] Yang Qingchun [杨庆春], "When Inspecting a Test-Oriented Training Base, Ma Xiaotian, Member of the Central Military Commission and Commander of the PLA Air Force, Emphasizes Accurately Determining and Deeply Understanding the Base's Functions and Tasks, Scientifically Planning and Comprehensively Consolidating the Foundation for Building and Development" ["准确定位深刻理解基地职能任务科学筹划全面夯实建设发展根"], *Kongjun Bao*, December 23, 2014, p. 1.

[57] Dong Bin [懂宾], "Courage and Character Shine on the Drill Grounds: Reporter Observations of Air Force Troops Relying on 'System of Systems' Engagement Exercises to Carry Out Intensive Fighting Spirit Training" ["血性在演兵场张:空军部队依托体系对抗演习靠站战斗精神强训之记者观察"], *Kongjun Bao*, June 10, 2015.

peacetime and needed to be inspired about their mission. Self-criticism of the training focused on "over-sea operations, weapons and equipment research, and the balance of safety and results," and assessment scores were to be "measured on the scale of actual combat requirements."[58] Other state-run media said that the concept was intended to "quicken the evolution of actual combat-oriented training and to elevate the Air Force's strategic strike capability across the board."[59]

Reflecting CMC support for the PLAAF's new training framework and over-water training, during an October 2014 inspection of a PLAAF command post, CMC Vice Chairman Fan commended the post for training in "serial exercises for systems-of-systems confrontations every year . . . , independent air combat and confrontations, contest-orientated situations on penetration and assaults, and training for valley, distant sea, super-low-altitude," and noted that "'Golden Helmet' and 'Golden Dart' had become popular brands."[60] He said that the "Air Force can be deployed across the entire country . . . and carry operational training in all directions." He also noted that PLAAF training "focused on the areas where they might fight battles and the kind of battles they might fight in the future to carry out training."[61] Clearly this includes the maritime domain.

New Maritime Combat Training for Cadets

In addition to unit-based combat-oriented training, the PLAAF has also sought to incorporate maritime themes into early training curriculum for new pilots at its PLA Air Force Aviation University in Jilin Province. The PLAAF conducted maritime flight training for the first time in September 2014 using a J-8 trainer with flight instructors.[62] The instructors practiced "formation flight, ultra-low altitude over water flight, and search and investigation," and future training will include "single aircraft tactical mobility [and] one-on-one air combat." [63] The over-water training was in the works for a year, roughly when the broader PLAAF began its uptick in over-water training, and reflects the school's mission of "training what is necessary for future warfare." In May 2015, the school "conducted its first live-fire training at sea," which brought the school "closer to [active] units" and improved training for "actual combat."[64] This included a focus on

---

[58] Xu Tongxuan, July 16, 2015, p. 3.

[59] "Military Report," November 11, 2014.

[60] Xiong Weiwei, October 27, 2014, p. 1.

[61] Xiong Weiwei, October 27, 2014, p. 1.

[62] "PLA JL-8 Trainer Aircraft Conducts Maritime Training for First Time" ["我军 JL-8 教练机首次进行海上训练低空掠过客轮"], China Military Online, September 19, 2014.

[63] "PLA JL-8 Trainer Aircraft Conducts Maritime Training for First Time," September 19, 2014.

[64] Li Kaiqiang [李开强] and Zhang Hongwei [张宏伟], "Air Force Organized First Live Fire Training at Sea for Flight School" ["空军首次组织飞行院校海上实弹打靶"], *PLA Daily*, May 20, 2015.

the impacts of maritime weather on strike accuracy and search and rescue missions. The school also created textbooks for students, including *Teaching Materials for Comprehensive Over Water Training with Trainer Aircraft*, *Syllabus for Comprehensive Over Water Training with Trainer Aircraft*, *Practical Handbook on Maritime Live-Fire Training with Trainer Aircraft,* and *Safety Checklist for Maritime Live-Fire Training.*[65]

Other early career training programs are also supporting the related mission of long-distance power projection. In April 2016, the Harbin Flight Academy, which trains PLAAF bomber pilots, restarted using recently retired H-6 bombers to train students for the first time in nine years, as Chinese media reported more new H-6 planes are entering service.[66] This will "substantially shorten the time required to train bomber crews and strongly boost the bomber force's development," according to a PLAAF statement quoted in Chinese media.[67] These efforts are clearly intended to support the long-term development of PLAAF pilots who can perform over-water missions.

Lessons Learned from Early Experimentation

A significant milestone for the PLAAF's development of over-water training occurred in September 2015 when the PLAAF held a two-day meeting—attended by senior PLA leaders, including PLAAF Commander Ma—that reviewed its performance in a PLAAF "maritime penetration and assault" contest and evaluation.[68] The meeting was intended as a capstone to collect early lessons and solutions for enhancing "aerial offensive operation capability" through the new maritime and combat-oriented focus on training, as Ma referenced Shenyang and Nanjing MR efforts on "design and employment of penetration tactics" and Jinan MR efforts on "assault training." During the session Ma stated that

> The purpose of holding this review and reflection meeting is to raise the combat-realistic level of the Air Force's military training and enhance the Air Force's operation capabilities in the maritime domain around the implementation of the Central Military Commission's military strategic guideline in the new period in accordance with the strategic requirement of more quickly building a powerful People's Air Force that is based on air and space integration and possesses capabilities of offense and defense.[69]

---

[65] Li Kaiqiang and Zhang Hongwei, May 20, 2015.

[66] Students previously used Y-7 transport planes. See Zhao Lei, "PLA Shifts Bombers to Flight-Training Role," *China Daily*, May 13, 2016; and Li Kaiqiang [李开强] and Wang Zhijia [王志佳], "Cadets Fly Bombers for First Time" ["空军飞行学员首飞喷气式轰炸机"], China Military Online, April 12, 2016.

[67] Li Kaiqiang and Wang Zhijia, May 20, 2016.

[68] Xu Tongxuan, September 17, 2015, p. 1.

[69] Xu Tongxuan, September 17, 2015, p. 1.

The main event was designed to bring together "command organs and combat units" with "factories and research institutes" to jointly study the major findings derived from the maritime penetration and assault contest. The after-action evaluation included a wide range of experts to study the problem's theoretical aspects and develop solutions to operational and technical problems exposed during the exercise.

PLA reporting on this experimentation indicates that PLAAF capabilities in performing its maritime support roles are in a relatively early stage, with dedicated units tasked to study key elements of operational concepts and tactics development remaining.[70] Indeed, Commander Ma stated that "the building of the air offensive operation capability in the Air Force remains in a transitional period of changing concepts, in a developing period of equipment used in offensive operations, and in the initial period of combat-realistic training and capability enhancement." Similar to other aspects tied to the PLAAF's modernization efforts, the development of operational capabilities remains in a transitional period. This phase of development is characterized by identifying, testing, and building new operational concepts that take advantage of the significant improvements in PLAAF technology and systems. Future exercises most likely will take the results derived from events such as this and use them as the basis for developing new tactics and building operational experience more broadly within the PLAAF.

[70] Xu Tongxuan, September 17, 2015, p. 1.

4. Recent Training for Operations over Water

In recent years, PLA missions beyond China's borders have expanded largely because PRC leaders adopted the "going global" strategy requiring that the PLAAF increase its participation in international affairs.[71] A notable example of this trend can be seen in the PLAAF's recent cases of "distant sea training." At the Zhuhai Air Show in November 2016, PLAAF spokesperson Senior Colonel Shen Jinke reviewed the PLAAF's modernization progress since President Xi came to power in 2012 and noted that "the PLA Air Force has flown over the First Island Chain (飞越岛链), carried out management and control of the East China Sea (管控东海) and combat patrols in the South China Sea (战巡南海) and developed high-tech weapons and equipment at a faster pace. The PLA Air Force's strategic transformation is moving from quantitative to qualitative changes."[72] Speaking again a month later, Shen further explained that "in the two years of distant sea training, the PLAAF had . . . engaged in training for reconnaissance and early warning, maritime patrolling, maritime assault, and mid-air refueling, among others, thereby improving distant sea mobility and testing distant sea combat ability."[73] Returning to a common refrain, Shen asserted that the flights improved the PLAAF's ability to "safeguard the country's sovereignty and territorial integrity as well as safeguard the nation's airspace security and development interests."[74] These accomplishments were made possible by the PLAAF's rapid increase in over-water training since 2015.

This training demonstrated a dramatic growth in long-distance maritime power-projection capabilities, especially through four groundbreaking flights into the Pacific Ocean through the First Island Chain in 2015 and flights into the South China Sea and around Taiwan in 2016. By the authors' count, since March 2015, the PLAAF has conducted eight flights past the First Island Chain, including three patrols of the East China Sea Air Defense Identification Zone (ADIZ) and two flights around Taiwan, as well as five flights into the South China Sea.

These flights have both operational value for training pilots and political value for strategic messaging. While these flights have been couched as normal operations for littoral states and

---

[71] Yu Hongwei [余泓伟] and Ren Bin [任斌], "China's Air Force Goes to Western Pacific for Distant Sea Training" ["中国空军赴西太远海训练 飞出第一岛链上千公里"], China News Online, August 14, 2015.

[72] Li Kaiqiang [李开强], "Chinese Air Force Spokesperson: PLA Will Continue Patrols in East China Sea ADIZ" ["空军发言人:解放军将继续在东海防空识别区警巡"], *PLA Daily*, November 5, 2016.

[73] Zhang Yuqing [张玉清] and Zhang Mimi [张汨汨], "China's Air Force Strengthens Distant Sea Training, Increases Strategic Capabilities" ["中国空军加强远海训练提升战略能力"], Xinhua, December 15, 2016.

[74] Chinese media translated this as: "safeguarding national sovereignty, security and development interests" (维护国家主权和领土完整，维护国家空天安全和发展利益). See "China's High Sea Military Drills Are Routine: Spokesperson," Xinhua, December 15, 2016; Zhang Yuqing and Zhang Mimi, December 15, 2016.

part of the natural development of the Chinese military in official PLAAF announcements, authoritative military commentary suggests that the utility of these flights extends beyond simply training for maritime missions, as they are sometimes also intended to convey strategic signals to relevant countries during times of political tension with China. There is also evidence that this training is laying the groundwork for strikes against regional targets, notably Guam.

The Southern TC appears to be taking the lead in experimentation for maritime activities.[75] A Southern TC bomber unit was revealed to have participated in the first flight into the Pacific in March 2015, the September 2015 maritime strike training, and the first flights into the South China Sea in 2016. Public reporting suggests that the Southern TC is primarily responsible for flights through the Bashi Channel and into the South China Sea, while the Eastern TC flies through the Miyako Strait into the Western Pacific (see Table 1 and Figures 1 and 2). This decision is likely based on each TC's main strategic focus and also on its having the closest units to each route.

Four Flights Past the First Island Chain into the Pacific Ocean in 2015

The PLAAF conducted four flights past the First Island Chain into the Pacific Ocean in 2015, two through the Bashi Channel in March and August, and two through the Miyako Strait in May and November. While the initial flights through each route were flown only by H-6K bombers, the follow-up flights were flown by a group of aircraft, reflecting increased training complexity.

Flight Details

In March 2015, a PLAAF unit overflew the Bashi Channel, a part of the Luzon Strait between Taiwan and the Philippines, on its way to over-water flight training. PLAAF spokesperson Shen said that this was "the first time [the PLAAF] conducted distant sea aviation training in the Western Pacific," and that it "improved the distant sea operational mobility of the Air Force [and was] an effective way…to temper its combat capability."[76] While no details were provided on location, we now know H-6K bombers from the Guangzhou MR did participate.[77] Foreshadowing the next phase of training, one Chinese military commentator speculated that in "future training, patrol, combat and carrier-borne early-warning aircraft will escort H-6Ks and

---

[75] Nathan Beauchamp-Mustafaga, Cristina Garafola, Astrid Cevallos, and Arthur Chan, "China Signals Resolve with Bomber Flights over the South China Sea," War on the Rocks, August 2, 2016.

[76] Yang Zhen [杨振] and Li Kaiqiang [李开强], "Chinese Air Force for the First Time Conducts Western Pacific Training" ["中国空军首次赴西太平洋开展远海训练"], *PLA Daily*, March 30, 2015.

[77] For details on earlier Guangdong MR H-6K over-water training and a possible March 2015 long-distance precision strike training, see Xu Sheng, Zhao Lingyu, and Wang Wenbin, December 8, 2015; Beauchamp-Mustafaga et al., August 2, 2016.

Table 1. Known Recent PLAAF H-6K Bomber Long-Range Operations over Water

Date of Operation (If Unknown, Date Information Was Released)	Region	Aircraft Involved	Description
March 2015 (announced)	Western Pacific	H-6K	Flight through Bashi Channel
May 2015 (announced)	Western Pacific	2 H-6K	Flight through Miyako Strait
August 2015 (announced)	Western Pacific	H-6K, other aircraft	Flight through Bashi Channel
September 2015 (announced)	Unclear, but described as several thousand km from southern China	H-6K	Training for long-range precision strikes
November 2015 (announced)	Western Pacific	4x H-6K, 1x Y-8, 1x Tu-154	Flight through Miyako Strait
November 2015 (announced)	East China Sea	4x H-6K, 1x Y-8 (and PLAAF reported fighters) (separate from above)	Show presence ("patrol") in East China Sea ADIZ
May 2016 (reported by Chinese media)	South China Sea	H-6K	Flight over Fiery Cross Reef
July 2016 (released by PLAAF social media)	South China Sea	2x H-6K	Flight over Scarborough Shoal
July 2016 (announced)	South China Sea	H-6K, J-11, possibly J-10, early warning aircraft and tankers	"Combat air patrol" over Mischief Reef, Woody Island, and likely Livock Reef
August 2016 (announced)	South China Sea	H-6K, Su-30, KJ-200, reconnaissance planes and tankers	"Combat air patrol" over Spratlys, Scarborough Shoal
September 2016 (announced)	Western Pacific	H-6K, Su-30, early warning aircraft, tankers, and other types of aircraft	Flight through Bashi Channel
September 2016 (announced)	Western Pacific	4x H-6K, Su-30, 1x Y-8, 1x Tu-154, (PLAAF reported tankers and early warning aircraft, possibly KJ-2000) (40 total reported, 8 confirmed)	Flight through Miyako Strait (Su-30s not confirmed to have transited Strait)
September 2016 (announced)	East China Sea	4x H-6K, 1x Y-8, 1x Tu-154 (and PLAAF reported fighters) (same planes as above)	Show presence ("patrol") in East China Sea ADIZ
November 2016 (acknowledged following media reports)	Taiwan	2x H-6K, 2x Su-30, 1x Y-8, 1x Tu-154	Circumnavigated Taiwan through Bashi Channel then Miyako Strait (Su-30s only transited Miyako Strait)
December 2016 (acknowledged following media reports)	South China Sea	H-6K and fighters	Followed "Nine-Dash Line" in South China Sea
December 2016 (acknowledged following media reports)	Taiwan	2x H-6K, 2x Su-30, J-10, 1x Y-8, 1x Tu-154	Circumnavigated Taiwan through Miyako Strait then Bashi Channel (Su-30s only transited Miyako Strait)
December 2016 (announced)	East China Sea	Unknown, likely planes from the above flight	Show presence ("patrol") in East China Sea ADIZ

SOURCES: PLAAF statements, Chinese state-run media, and Japanese MOD. Adapted from Beauchamp-Mustafaga et al., 2016.

NOTES: Number and type of aircraft are unknown unless specified. The Y-8 and KJ-200 are early warning aircraft, the Tu-154 is a reconnaissance plane, and the J-11 and Su-30 are fighters. Flights through the East China Sea ADIZ in November 2015 and September 2016 were likely part of larger formations that flew the same day near Japan, and both instances are listed above as flights in the Western Pacific. The December 2016 ADIZ flight is assumed to have occurred under similar circumstances. Neither the September nor December 2016 ADIZ flights made noticeable detours while on their way to fly through the Miyako Strait to constitute a patrol similar to the first one conducted in November 2015, according to Japanese MOD descriptions of their flight paths.

Figure 1. Known Recent PLAAF Bomber Long-Range Operations over Water

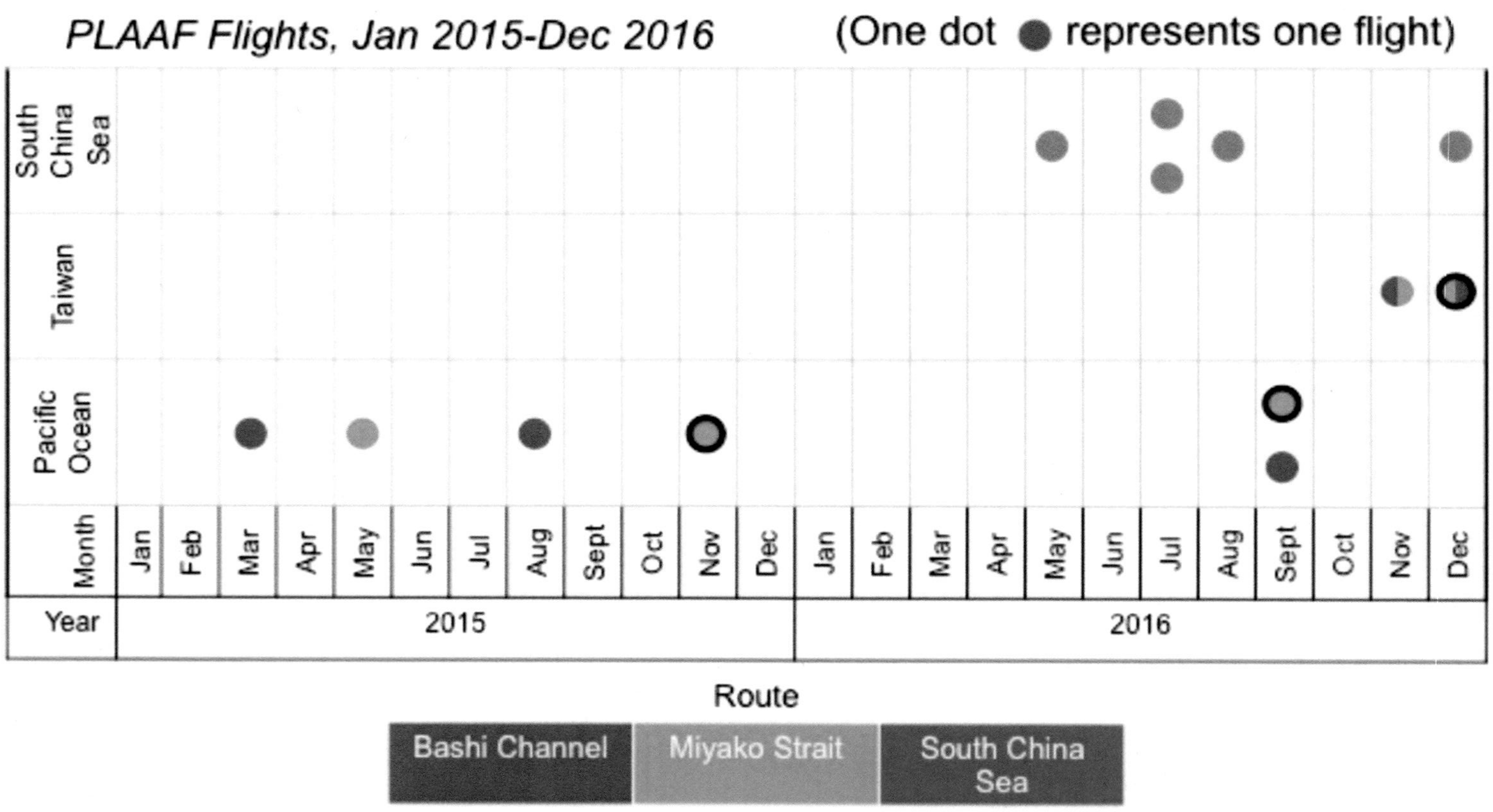

NOTES: Bolded circles indicate PLAAF patrols of the East China Sea ADIZ. The November 2015 ADIZ patrol is assumed to have been conducted by separate PLAAF H-6Ks, but the September and December 2016 ADIZ patrols are likely to have been the same planes that transited the Miyako Strait. Circles representing November and December 2016 flights around Taiwan are two different colors to indicate that planes flew through both Bashi Channel and Miyako Strait. Order of transit is indicated from left to right.

Figure 2. Map of Asia-Pacific

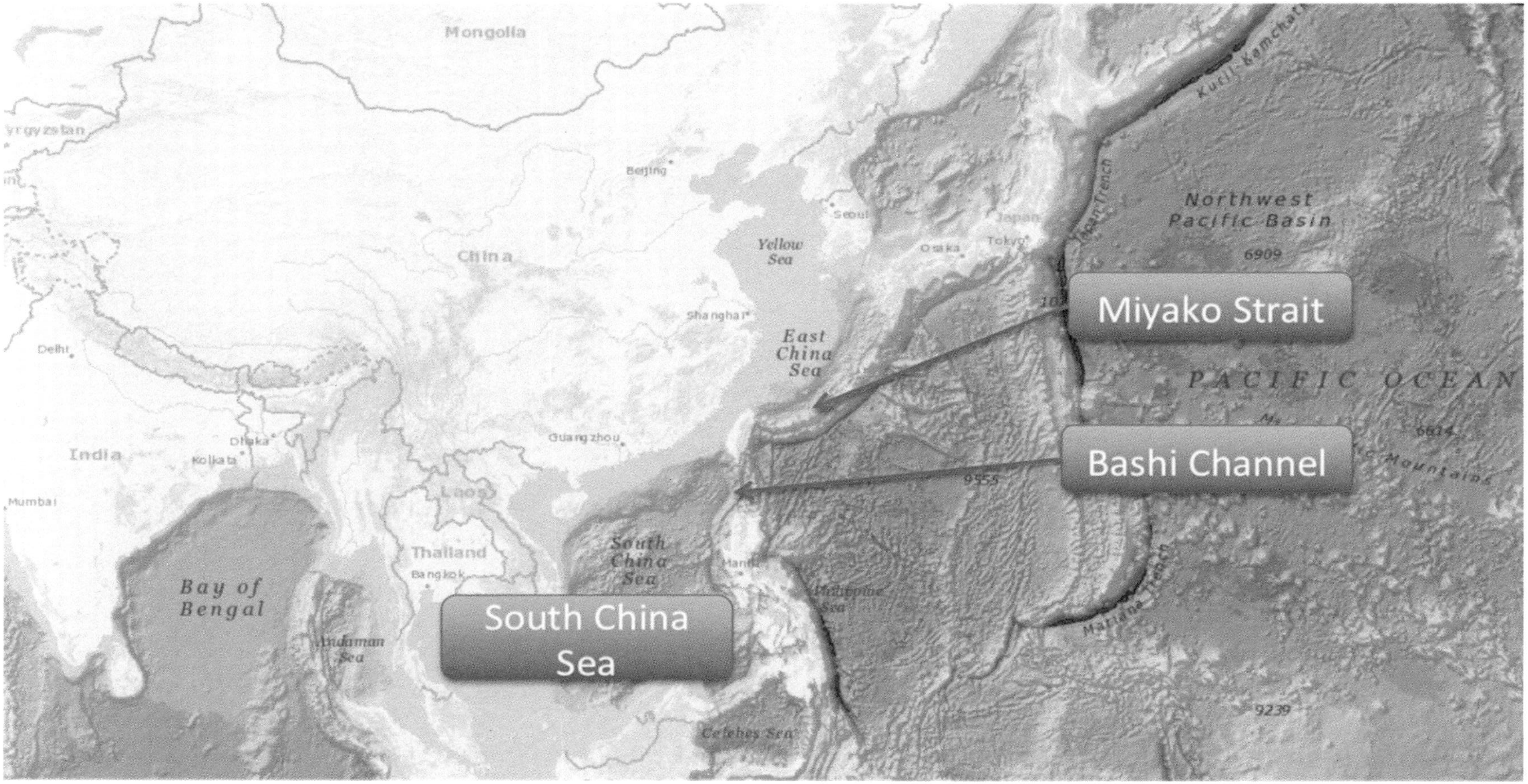

SOURCE: Google Maps.

provide overflight and a counter-electromagnetic environment. HZ-6 (reconnaissance/bombers), HY-6 (aerial refueling tankers) and early warning aircraft may also participate in the training."[78]

Subsequently, during the PLAAF's second flight through the Bashi Channel (and third overall long-distance flight into the Pacific) in August 2015, several types of aircraft, including the H-6K, flew through the Channel on their way 1,000 km beyond the First Island Chain.[79] It appears from photos in state-run media that at least four H-6K bombers participated in the flight. Shen framed the flight as "testing [the PLAAF's] study power and combat power," since it is a "strategic service . . . facing new historic missions." The H-6K is described by the PLA as "one of the iconic equipment items in the transformation of the PLA Air Force into a strategic air force," and its progress for "combat power" is measured by "'extending the distance of flight' and "dealing accurate strikes."[80]

In May 2015, during the second of the four ground-breaking flights that year, two H-6Ks overflew the Miyako Strait near Okinawa, Japan, and "conducted open sea training in the western Pacific."[81] This was the first time PLAAF planes had flown through the Strait on their way to the Pacific and was the second overall PLAAF flight past the First Island Chain. Of note, Japanese MOD reporting reveals that two H-6s flew through the Miyako Strait for the first time in September 2013 and then again in October 2013, March 2014, and December 2014.[82] Although the Japanese government did not identify the service, they appear to have been PLAN

---

[78] Qiu Yue [邱越], "Chinese Air Force Distant Sea Training Must Be Regularized, Don't Rule Out Possibility of Foreign Military Surveillance" ["中国空军远海训练必将常态化 不排除遭外军监视可能"], People's Daily Online, March 31, 2015.

[79] Zhang Tao, "Expert: H-6K Demonstrates Long-Range Strike Capability," China Military Online, November 4, 2015; Qiu Yue [邱越] and Xiao Hong [肖红], "Several Types of Chinese Air Force Fighters Passed 1,000 km Beyond the First Island Chain to Conduct Distant Sea Training" ["中国空军多型战机飞出第一岛链 1000 余公里展开远海训练"], People's Daily Online, August 14, 2015; and Wang Dongming [王东明], Zhang Yuqing [张玉清], and Wang Nannan [王楠楠], "What All Did the Chinese Air Force Do This Year: Flew Through the First Island Chain Four Times for Training in the Western Pacific" ["中国空军今年都干了啥:4 次飞出第一岛链演兵西太"], Xinhua, December 29, 2015.

[80] Xu Sheng, Zhao Lingyu, and Wang Wenbin, December 8, 2015.

[81] For details on the flight, see Japanese Ministry of Defense, "On the Flight of Chinese Aircraft in the East China Sea" ["中国機の東シナ海における飛行について"], May 21, 2015; and Qi Bingxin [齐冰昕], "China's Air Force Conducts Distant Sea Training Through Miyako Strait for First Time" ["中国空军首次飞越宫古海峡开展远海训练"], China Air Force Online, May 22, 2015.

[82] The Japanese government also said, "in September 2007, multiple H-6 bombers flew into Japan's Air Defense Identification Zone (ADIZ) above the East China Sea and advanced near the Japan-China median line." See Japanese Ministry of Defense, "2014 Defense of Japan," 2014; and Japanese Ministry of Defense, "2016 Defense of Japan," August 2, 2016.

Air Force planes.[83] The May 2015 flight was framed as part of the PLAAF's "strategic transformation" to "fly farther, strike more precisely, and build stronger combat capabilities."[84]

In November 2015, PLAAF aircraft again overflew the Miyako Strait and continued 1,000 km past the First Island Chain.[85] Shen said, "several models, including H-6K bombers" were involved in the patrol, and Japanese MOD reporting reveals that the formation included four H-6Ks, one Y-8 early warning aircraft, and one Tu-154 reconnaissance plane. This reflects a similar training progression to incorporate more airframes, as in the Guangzhou MR, although there is no information available on specific airframes used by the Guangzhou MR for its March and August flights. At the same time, more H-6K and other undisclosed fighters and early warning aircraft also conducted a "patrol" of the East China Sea ADIZ, according to Shen.

Japanese MOD reporting reveals that the ADIZ patrol consisted of four H-6Ks and one Y-8, notably contradicting the PLAAF statement that the patrol included fighters.[86] Only the four H-6Ks actually deviated significantly from the original flight path and flew through the ADIZ closer to the Japanese main islands before turning around, whereas the Y-8 appears to have flown in small circles generally near the original flight path of the group that flew through the Strait. This suggests that a more apt description of the patrol would be a "show of presence" flight, since bombers do not typically conduct such patrols. Subsequent "patrols" of the ADIZ in September and December 2016 would not feature H-6Ks deviating significantly from the most direct route through the Strait. One military expert said that the "joint drills of H-6K bombers along with various types of jet fighters and airborne early warning surveillance aircraft" were notable because they were "different from the earlier training where only H-6K bombers were involved."[87] *IHS Jane's 360* identified at least some of these H-6Ks as from the Nanjing MR, which now makes up mostly the Eastern TC, and commented, "in the future more capable [early

---

[83] For increased PLA activity around Miyako as strategic signaling to Japan in fall 2013, see Peter Wood, "China-Japan Relations Worsen as Chinese Naval Aviation Flies Through 'First Island Chain,'" *China Brief*, November 7, 2013.

[84] Lin Yuan [林 源], Yu Hongchun [余红春], Yan Guoyou [闫国有], Chi Yuguang [迟玉光], Bai Zhongbao [白仲宝], and Mu Xingguang [牟星光], "Bombers Take Action" ["轰炸机出动"], *China Air Force* [中国空军], August 19, 2015.

[85] For details on the flight, see Japanese Ministry of Defense, "On the Flight of Chinese Aircraft in the East China Sea" ["中国機の東シナ海における飛行について"], November 27, 2015; and Zhang Yuqing [张玉清] and Wu Yi [伍轶], "Chinese Air Force's Planes of Several Models Including H-6K Go to Western Pacific Ocean for Blue Water Training; H-6K, Fighter Planes and Early Warning Planes of Several Models Simultaneously Dispatched to Conduct Policing Patrols in East China Sea Air Defense Identification Zone" ["中国空军轰-6K 等多型飞机赴西太平洋远海训练 同时出动轰-6K 和多型歼击机预警机警巡东海防空识别区"], Xinhua, November 27, 2015.

[86] For more details on the November flights, see Richard Fisher, "Chinese Bomber Exercise Affirms Air-Defence Identification Zone, Penetrates Second Island Chain," *IHS Jane's 360*, December 1, 2015; and Japanese Ministry of Defense, August 2, 2016.

[87] "Focus Today," China Central Television, November 30, 2015.

warning] platforms like the KJ-2000 or KJ-500 could provide an offensive search capability for attacking H-6K bombers."[88]

However, we cannot confirm the number or location of these flights, though it is unlikely that the PLAAF is underreporting them, whereas the PLAN Air Force appears less forthcoming about its similar flights. PLAN Air Force H-6 flights through the Miyako Strait in 2013 and 2014 were not as fully reported in Chinese media, judging by flights reported in Japanese Defense White Papers.[89] While the 2016 Japanese Defense White Paper reports additional flights in July over two days, consisting of two H-6s, one Y-9 reconnaissance plane, and one Y-8 early warning aircraft, it is likely that this is another unannounced PLAN Air Force flight, and as such has been left out of the tabulation above.[90] Other known PLAN Air Force flights match up with Japanese government reporting when the PLAAF did not announce or later acknowledge a flight through the Strait. Taken together, other foreign government reporting suggests that the PLAAF is generally forthcoming about these flights, though it may not detail specific flight paths for specific aircraft.

Strategic Messaging

PLAAF signaling on these flights focused on demonstrating China's growing bomber capabilities and ability to enforce Chinese territorial claims. PLAAF spokesperson Shen said that as a strategic service, the PLAAF's "strategic capability should correspond to national interests to provide reliable air and space security for national development," and the four flights into the Pacific in 2015 "[boosted] the capability of the airmen for long-range, blue-water 'system-of-systems' operations."[91] Professor Li Li from China's National Defense University said that these flights "[indicated] that the formulation and maintenance of rules over the East China Sea ADIZ have been in smooth progress" and that "we are not only able to safeguard the East China Sea ADIZ, but also fly beyond the First Island Chain."[92] Professor Li added that the "dispatch of the bomber fleet is the symbol of a strong air force." Another military expert said that the four flights "showed the presence of the PLA's airpower in regions that are closely related to Chinese national security," "China's conventional military deterrence capability," and the H-6K's "powerful conventional strike capability."[93] Another said, "new-type Chinese bombers taking

---

[88] Fisher, December 1, 2015.

[89] A few Chinese media outlets did report on the initial September 2013 flight, with Chinese media citing Japanese media reports, but not all later flights were reported. See "H-6 Bomber Flies over Japan's Southwestern Islands and Enters for First Time" ["轰 6 轰炸机首次飞越日本西南诸岛进入太平洋"], *Observer* [观察者], September 8, 2013; Japanese Ministry of Defense, August 2, 2016.

[90] Japanese Ministry of Defense, August 2, 2016.

[91] Zhang Yuqing and Wu Yi, November 27, 2015.

[92] Zhang Tao, "Expert: China Capable of Defending East China Sea ADIZ," China Military Online, December 2, 2015.

[93] Zhang Tao, November 4, 2015.

part in this offshore military drill have signaled a marked progress of China's long-range strategic delivery capability."[94]

These flights appear to reflect early PLAAF efforts to normalize long-distance training into the Pacific and desensitize regional countries to the activities. PLAAF spokesperson Shen's comments following several flights emphasized that these were normal exercises, similar to ones conducted by other countries, and that flights were "routine arrangements of the annual training plan" and thus would continue.[95] He also noted that the flights were "not aimed at any country or target and pose no threat to any country or region."[96] These remarks were repeated for many, if not all, subsequent flights. The Miyako Strait was described as "an international waterway" where "all countries are entitled to the freedom of navigation and over-flight."[97] In comments carried in PLA media, military law expert Xing Hongbo said, "other countries are expected to understand, respect and adapt to the drills by the Chinese Air Force beyond the First Island Chain," and elsewhere Xing wrote that "distant sea training [was] legal, legitimate [and] justifiable," a phrase that has been repeated frequently by Shen when discussing H-6K flights.[98] Another military expert wrote that since "China…is not yet a strong maritime power [and] was bullied and humiliated because of its weak sea defense" in the past, "now it is quite necessary for the Chinese Navy and Air Force to raise their maritime defensive combat capability through distant sea training, which is an objective requirement for safeguarding the country's maritime security."[99] He added that international calls for China to "assume more international responsibilities" means that "the Chinese Navy and Air Force [need] to raise their distant sea activity capability." PLAAF efforts to frame the flights as typical military training suggest that the PLAAF intends the flights to function partially as deterrence signaling but also does not want to provoke regional tensions every time it goes into the Pacific.

Chinese military commentary reflected official rhetoric that flights would continue and suggested potential new routes into the Pacific. One military commentator, speaking after the first flight through the Bashi Channel in March 2015, said, "In the future, blue-water flight training of the Chinese Air Force will surely become an ordinary training evolution and cooperative blue-water training with the navy will be conducted."[100] The expert added that the

---

[94] "From Phoenix to the World," Phoenix TV, May 21, 2015.

[95] Yao Jianing, "PLA Air Force Conducts First Training in West Pacific," China Military Online, March 30, 2015.

[96] Yao Jianing, March 30, 2015.

[97] Guo Renjie, "PLA Air Force's Flying over Miyako Strait for Training: Legitimate and Reasonable," China Military Online, May 22, 2015.

[98] Guo Renjie, May 22, 2015; and Xing Hongbo [邢洪波],"The PLA Air Force's Distant Sea Training Is Legal, Legitimate, Justifiable" ["中国空军飞越宫古海峡赴远海训练合法合理合情"], People's Daily Online, May 22, 2015.

[99] Zhang Junshe [张军社], "Japan Needs to Get Used to Chinese Air Force's Distant Sea Training" ["日本应习惯中国空军远海训练"], *People's Daily Overseas Edition* [人民日报海外版], April 1, 2015, p. 1.

[100] Qiu Yue, March 31, 2015.

PLAAF may also transit the Osumi Strait, between Japan's Kyushu and Okinawa Islands, in the future, hinting that it could be the Shenyang MR, which is now mostly the Northern TC. As of December 2016, PLAAF planes were not known to have flown through the Osumi Strait or to have conducted joint training with the PLAN past the First Island Chain.

Media reporting suggests that the Guangzhou MR, now the Southern TC, as the first military region to field the H-6K, is leading the PLAAF's experimentation with the new bomber, especially for over-water operations. According to one *PLA Daily* article,

> With the H-6K bombers in active service, the [Guangzhou MR] unit has scored outstanding and exciting achievements: In the normalized distant sea training, and in performing more than 30 major tasks, including penetration and assault drills, the H-6K bombers achieved impressive results again and again.

It adds,

> In the long-duration distant sea flight training organized by the Air Force for the first time, they created multiple records in the PLA Air Force, such as the longest distance of flight above the sea, the longest duration of continuous flight, and the longest range of operation flight. When taking part in targeting practice with actual missiles, by surmounting difficulties, the unit operated under a marginal and extreme condition, and achieved excellent scores.[101]

Indeed, Liu Rui, a bomber pilot with the Southern TC, was the first to fly past the First Island Chain in March 2015, flew in the military parade in September 2015, and was the first to fly in the South China Sea in 2016, reflecting the TC's leading role.[102]

"Combat Air Patrols" over the South China Sea in 2016

The PLAAF appears to have shifted its over-water power-projection focus to the South China Sea in the summer of 2016, likely in a political response to the Permanent Court of Arbitration's (PCA) ruling against China in July.[103] The PLAAF announced two H-6K bomber flights over the Spratly Islands, and Chinese state-run media reported two other flights, suggesting at least four flights into the South China Sea between May and August 2016 covering both the Spratly and Paracel Islands.[104] This was the first acknowledged PLAAF presence in the South China Sea, as previously identified aircraft were all PLAN Air Force. Moreover, in July 2016, PLAAF spokesperson Shen said that the patrols would be "regularized" and continue, and a military

---

[101] Xu Sheng [徐生] and Yu Hongwei [余泓纬], "A Unit of the Guangzhou Military Region Air Force Forges All-Weather, Whole-Territory Combat Capability with New-Model Strategic Bombers" ["广空某部全天候全疆域锤炼新型战略轰炸机作战能力: 战鹰振翅苦练硬功显威蓝天"], *PLA Daily*, January 3, 2016, p. 1

[102] "Exclusive Revelations: 'God of War' H-6K Patrol of Scarborough Shoal" ["独家揭秘"战神轰炸机"轰-6K 巡航黄岩岛"], China Central Television, November 5, 2016.

[103] Jane Perlez, "Tribunal Rejects Beijing's Claims in South China Sea," *New York Times*, July 12, 2016.

[104] Beauchamp-Mustafaga et al., August 2, 2016.

commentator said that they would cover more of the South China Sea. Of note, Chinese H-6 bombers operated in a similar manner in the South China Sea in the early 1980s, according to declassified U.S. documents, and the Chinese media reported that the PLAAF "had been conducting 'routine surveillance' in the area since 2012."[105] In contrast with flights into the Western Pacific, the PLAAF did not claim that these flights into the South China Sea were a first for the service, suggesting some reliability for PLA sources.

Flight Details

The first public appearance of the H-6K over the South China Sea was in early May, when China Central Television aired a show about an H-6K pilot and included a short clip of the plane flying over Fiery Cross Reef in the Spratlys.[106] In mid-July, shortly after the ruling by the PCA against China, the PLAAF released photos of an H-6K flying over Scarborough Shoal, marking the second unannounced flight.[107] Several days later, the PLAAF made the first official announcement of H-6K flights in the area, calling them "combat air patrols" and saying, "the PLA sent H-6K bombers and other aircraft including fighters, scouts and tankers to patrol islands and reefs including [Scarborough Shoal]" and that they had accomplished the objective of the combat air patrol by conducting "aerial scouting, air combat and island and reef patrol." [108] PLAAF spokesperson Shen added that the patrols would become "regularized" and the flight had deepened the "combat realistic training in the maritime direction." Chinese media later confirmed that the fighters included J-11 aircraft and likely J-10 aircraft.[109] Footage aired on China Central Television included land features identified as Mischief Reef and Livock Reef in the Spratlys and Woody Island in the Paracels, the H-6K's first appearance in the northern part

---

[105] Only the PLAN Air Force is known to have conducted flights in the South China Sea in the 1980s. A declassified Central Intelligence Agency (CIA) report from 1984 says,

> On 8 November 1980, two naval B-6 bombers from Hainan Island overflew Northeast Cay, the northernmost island of the Spratlys. In anticipation of Vietnamese reaction, Chinese fighter aircraft conducted defensive patrols near the Paracels while the bombers flew south…In October 1983, two B-6 bombers overflew Malaysian-occupied Swallow Reef in the Spratlys, this time with a frigate and replenishment oiler deploying south of the Paracels to provide navigational assistance and air surveillance.

See National Archive CIA Records Search Tool, *China: Military Options Against Vietnam*, March 1, 1984, Declassified CIA-RDP84S00928R0003000050006-0, p. 10, cited in Peter Wood, "Snapshot: China's Southern Theater Command," *China Brief*, July 22, 2016; see also Kinlin Lo, "'It Was a Provocative Act': PLA Pilot Describes Encounter with Foreign Military Jet," *South China Morning Post*, February 13, 2017.

[106] "My Story with the God of War," China Central Television, May 6, 2016; and Richard Fisher, "Chinese H-6K Bomber Flies over Fiery Cross Reef in South China Sea," *IHS Jane's 360*, May 12, 2016.

[107] PLAAF Weibo account, July 15, 2016; and Heng Jing [靖恒], "Chinese Air Force for the First Time Releases Photos of H-6K Patrolling Scarborough Shoal" ["中国空军首次曝光轰-6K巡航黄岩岛高清照片"], *Observer*, July 15, 2016.

[108] "China's Air Force Conducts Combat Air Patrol in South China Sea," Xinhua, July 18, 2016.

[109] Zhang Tao, August 25, 2016.

of the South China Sea.[110] It is unclear if the photos released in mid-July were from the same flight reported by Shen or from earlier, separate flights. A PLAAF pilot later claimed in an interview with Chinese state-run media that during one of these July flights, he encountered a foreign fighter jet over the South China Sea.[111]

H-6Ks made a second official flight in early August, this time over Scarborough Shoal and other unidentified Spratly land features. The flight also involved "Su-30 fighters . . . Airborne Early Warning Aircraft [identified as KJ-200 in Chinese media], reconnaissance and tanker airplanes," and the Su-30s refueled twice.[112] Shen described the flight as "establishing a combat system integrating air control over the sea, defense penetration, surprise attack, support and supply."[113] Of note, the Su-30 fighter was the first official identification of other planes flying with the H-6K, as they are usually referred to generically, and Chinese media said that the early warning aircraft was the KJ-200.[114] Also, this flight appears to have included additional training elements beyond the July flight, as one military commentator said, "the combat patrol included air defense early warning maneuvers, air combat confrontation, and island patrolling in a complicated electromagnetic environment," of which the early warning and complex electromagnetic environment were new.[115]

Several months later, in early December, H-6K bombers and unknown fighters conducted another flight through the South China Sea. This was first reported by U.S. media and later confirmed by Shen with no further details.[116] U.S. officials cited in U.S. media reports said that the H-6K "flew along the disputed 'Nine-Dash line,'" which represents China's territorial claims in the region, and flew farther on this flight than on previous flights during the summer, without providing more details of the flight path. The officials also said that it was "the first long-range flight of a Chinese bomber along the U-shaped line of demarcation since March 2015," although

---

[110] For additional details on the flight, see "Exclusive Revelations: . . . ," November 5, 2016; "Chinese Air Force H-6Ks Conduct Combat Air Patrol over South China Sea and Scarborough Shoal" ["中国空军轰—国空战斗巡航南海黄岩岛"], China Central Television, July 18, 2016; and Beauchamp-Mustafaga et al., August 2, 2016.

[111] The same pilot claimed in December 2016 that he encountered foreign fighters on his first distant seas flight, which is likely the March 2015 flight through the Bashi Channel into the Western Pacific. See Wang Tonghua [王通化] and Li Kaiqiang [李开强], "For the Country, a Sharp Sword" ["为国仗剑'锐一代'"], *PLA Daily*, December 29, 2016; Lo, February 13, 2017; and "One on One: Liu Rui: The 'God of War' Story" ["[面对面]刘锐：'战神'故事"], China Central Television, February 12, 2017.

[112] "Chinese Military Aircraft Patrol South China Sea," Xinhua, August 6, 2016; and "Chinese Air Force Combat Aircraft Conduct 'Combat Air Patrol' over South China Sea" ["中国空军多型主战飞机赴南海战斗巡航"], China Central Television, August 6, 2016.

[113] Zhang Tao, August 25, 2016.

[114] Zhang Tao, August 25, 2016.

[115] "Focus Today," China Central Television, August 7, 2016.

[116] Tomlinson, December 9, 2016; Zhang Zhaoying [张骄瀛], "Expert: China's Air Force Distant Sea Training Has Major Transformation, Air Combat System Already Formed" ["专家:中国空军远海训练有重大变化 空战体系已形成"], People's Daily Online, December 16, 2016.

there were no previously reported PLAAF flights through the South China Sea in 2015, and they may have been confusing it with the first long-range PLAAF flight through the Bashi Channel.

All these flights into the South China Sea appear to have come from the Southern TC, and at least some flights can be associated directly with the 8th Bomber Division headquartered in Leiyang, Hunan Province.[117] Although one Chinese analysis suggested that some of these flights contained in China Central Television footage took place between January and March 2016, judging by the progress of Chinese construction on reclaimed land, these flights in total encompass at least a seven-month period and five land features across a wide swath of the South China Sea.[118]

Strategic Messaging

The flights into the South China Sea appear intended as strategic conventional deterrence signaling to reinforce China's willingness to defend its claims to disputed land features and demonstrate that its operational range covers all of China's claimed territory, likely in response to the PCA ruling against China. Although Chinese officials did not explicitly link the flights with the PCA ruling, Chinese state-run media made the connection clear.

During a visit to a PLAAF base in the Southern TC in mid-July, CMC Vice Chairman Fan, accompanied by PLAAF Commander Ma and other senior military leaders, reviewed H-6Ks and said that the military must "continuously increase its deterrent and warfighting capabilities" and be able to fight and win wars.[119] Announcing the first official flight, PLAAF spokesperson Shen said, "The PLA Air Force will firmly defend national sovereignty, security and maritime interests, safeguard regional peace and stability, and cope with various threats and challenges."[120] In the first Chinese MND briefing following the ruling, when asked about the H-6K patrols, spokesperson Senior Colonel Yang Yujun said, "China has sovereignty over the islands and reefs of the South China Sea and their adjacent waters. It is the sure responsibility of the Chinese military to safeguard the national sovereignty security, territorial integrity as well as maritime interests and rights."[121] Drawing a more explicit connection, one military news article said that the flights "had something to do with the illegal South China Sea arbitration unilaterally

---

[117] Beauchamp-Mustafaga et al., August 2, 2016.

[118] "J-11 Escort H-6K Fly over Mischief Reef" ["歼-11 护航轰-6K 飞越美济礁 镜头长时间对准中业岛"], Phoenix TV, July 20, 2016.

[119] General Fan also visited the Spratlys in April 2016, see Chris Buckley, "Chinese General Visits Disputed Spratly Islands in South China Sea," *New York Times*, April 15, 2016; Minnie Chan, "PLA Unveils New Weapons for Air and Sea Combat Following Hague Tribunal Ruling," *South China Morning Post*, July 22, 2016; and "Fan Changlong . . . ," July 19, 2016.

[120] Dong Zhaohui, "PLA Air Force Conducts Combat Air Patrol in South China Sea," Xinhua, July 19, 2016.

[121] Zhang Tao, August 25, 2016.

initiated by the Philippines and also demonstrated the Chinese Air Force's stance that it is able to conduct effective control over the island."[122]

The main PLA commentator on PLAAF activities in the South China Sea, Senior Colonel Du Wenlong, has reinforced the importance of H-6K flights for Chinese deterrence and their ability to strike anywhere within Chinese territory. Du said, "with the aerial refueling capability, the PLAAF is making significant progress in improving combat support capabilities, extending the combat radius of its aircraft, and increasing its air strike capability when carrying out distant sea operations," specifying that "with the Su-30 bomber, China's air patrols can reach James Shoal, the southernmost territory of China."[123] He also noted that "the [early warning aircraft] plays an important role in increasing the PLAAF's capabilities to conduct distant sea operations with accurate detection of targets, commands of precision strikes, and better support system." Placing the flights in the PLAAF's overall transition to maritime missions, Du said, "the PLAAF used to focus on land strikes and did not make islands, reefs, and ships as actual targets, but now it has to be able to strike them all."[124] Du touted three features of the H-6K: long range to cover all of "the region's strategic targets with a devastating strike;" it can deter or strike "large formations" in the water that were "provoking China" if it has the supersonic YJ-12 anti-ship cruise missile (ASCM); and the H-6K's endurance to stay in the air, allowing it to cover the South China Sea like a net.[125] Looking to the future, Du said, "future patrols in the South China Sea will go beyond [Scarborough Shoal] to cover all islands and reefs within China's territory."[126]

Other military analysts placed the H-6K in the large PLAAF formations that flew and touted the opportunity for closer joint service training with the PLAN Air Force. Another military commentator added that "bombers, fighter jets, [early warning aircraft], and reconnaissance aircraft form a complete combat system to conduct patrol missions [and] the substantial extension of China's air patrol areas play a key role in safeguarding China's sovereignty."[127] A third commentator noted that the "regular air patrols will enhance its joint-force combat

---

[122] Zhang Tao, August 25, 2016.

[123] "Focus Today," August 7, 2016.

[124] "Asia Today," China Central Television, July 16, 2016; and Qiu Yue [邱越], "Expert: Chinese Air Force Backbone Can Strike Entire South China Sea, Future Patrols Will Visit More Islands and Reefs" ["专家:中国空军长臂可辐射整个南海 未来将巡航更多岛礁"], People's Daily Online, July 18, 2016.

[125] It is unclear why Du mentioned a possible ASCM capability for the H-6K, because it is was not known to have such a capability as of 2016. See "XAC H-6"; Qiu Yue [邱越], "Air Force Combat Air Patrols in the South China Sea to Be Regularized, Expert: China Has the Capability to Safeguard Sovereignty" ["空军南海战斗巡航将常态化 专家:中国有能力维护主权"], People's Daily Online, July 19, 2016.

This claim has been repeated in other nonauthoritative media. "Multiple Types of Chinese Air Force Aircraft Coordinate Defense Penetration and Assault: H-6 and Su-30 [Conduct] Anti-Carrier [Operations]" ["中国空军多机种协同突击突防: 轰 6 苏 30 反航母"], Sina, October 30, 2016.

[126] Zhang Tao, August 25, 2016.

[127] "Focus Today," August 7, 2016.

capabilities and symbolize that the PLAAF and Navy are capable of conducting distant sea operations."[128] Retired Colonel Yue Gang said that the PLAAF was also responsible for "[offering] protection to the naval forces in times of war."[129]

Another military commentator said that the use of dissimilar aircraft showed progress for the PLAAF's ability to coordinate takeoffs from different airfields and form into a group, which is "a tremendous challenge for the Chinese Air Force's command dispatch capabilities."[130] This commentator also focused on the importance of electromagnetic dominance in air combat, saying "in the future the South China Sea quite possibly may be filled with complex electromagnetic signals from four dimensions including from space satellites, combat aircraft in the air, warships at sea, and ground-based radar." The acknowledgment of difficulties for aircraft coordination marks a distinct shortcoming compared with the power-projection capabilities of the U.S. Air Force, which has been conducting long-distance bomber operations over water since the 1940s.

There also appears to be some internal debate over how overt the signaling should be. Reuters reported in July that "options floated by those linked to the PLA include putting missiles on bombers patrolling the South China Sea capable of hitting targets in the Philippines or Vietnam," and Du said, "Chinese aircraft will carry live ammunition in the patrol mission," though this has not been confirmed by official PLAAF statements.[131] Several commentators said that H-6Ks would continue to operate as part of the "combat air patrols" in the South China Sea.

Last, the most recent December 2016 flight was described by Senior Captain Liang Fang as signaling three messages. First, "that China maintains its sovereignty and jurisdictional claims within the nine-dash line; [second] that China was unafraid of threats in the South and East China Seas and capable of winning military clashes there; [and third] that President-elect Trump should not play with fire over Taiwan and, likewise, neither should the president of Taiwan."[132]

However, the H-6Ks are unlikely to feature prominently in routine patrols of a potential future ADIZ, as bombers rarely perform patrol functions. Moreover, despite the runway on Fiery Cross Reef being long enough to support a bomber and construction of hardened aircraft shelters on some reclaimed land features big enough for a bomber, operating conditions in the ocean environment would likely preclude long-term basing in the South China Sea.[133]

---

[128] "Focus Today," China Central Television, July 19, 2016.

[129] Catherine Wong, "PLA Air Force Holds Fresh Patrols in South China Sea as It Seeks to 'Normalise' Stepped-Up Surveillance," *South China Morning Post*, August 7, 2016.

[130] Ma Jun [马俊], "Dissimilar Main Force Combat Aircraft Combat Patrols of South China Sea: Chinese Air Force Shows Its Strength" ["多型主力战机战斗巡航南海!中国空军展示实力"], *Global Times*, August 8, 2016.

[131] Ben Blanchard and Benjamin Kang Lim, "'Give Them a Bloody Nose': Xi Pressed for Stronger South China Sea Response," Reuters, July 31, 2016.

[132] Steven Stashwick, "US Media Over-Hypes Chinese Bomber 'Warning' in South China Sea," *The Diplomat,* December 15, 2016; and Liang Fang [梁芳], Post on Sina Weibo [新浪微博], December 12, 2016.

[133] David E. Sanger and Rick Gladestone, "New Photos Cast Doubt on China's Vow Not to Militarize Disputed Islands," *New York Times*, August 9, 2016; Beauchamp-Mustafaga et al., August 2, 2016.

Flights into the Pacific Ocean in 2016

The PLAAF conducted another four flights past the First Island Chain in 2016. These flights began in September, later than in 2015, but with much quicker tempo and also utilizing new flight paths. Two flights were directly through the First Island Chain into the Western Pacific, one each through the Miyako Strait and Bashi Channel, similar to the 2015 flights. Moreover, PLAAF spokesperson Shen in September said that these flights through the First Island Chain would be "regularized."[134] Two more flights that also flew through the First Island Chain and then circumnavigated Taiwan are described in the next section. Chinese commentary regarding these flights suggests that the political intention was still to convey strategic messages about enforcing Chinese territorial claims and defending Chinese territory.

Flight Details

On September 12, 2016, a group of PLAAF H-6K bombers, Su-30 fighters, early warning, refueling, and other types of aircraft flew through the Bashi Channel into the Pacific for distant sea training, according to PLAAF spokesperson Shen.[135] They practiced reconnaissance and early warning, maritime patrolling (海上巡航), and mid-air refueling, among other skills. Most important, Shen announced that the PLAAF would "regularize" distant sea training past the First Island Chain, after again explaining that the flights were normal activities for littoral states and were part of the PLAAF's yearly pre-scheduled training. He also said that the PLAAF would "fly past the island chains, control the East China Sea, conduct combat readiness patrols in the South China Sea, and sharpen its sword in the plateaus," reflecting an emphasis on the PLAAF's maritime missions. One foreign analyst noted, "The presence of refueling tankers is curious since both the H-6K and Su-30 have ranges in excess of 1,000 miles, meaning they could easily make the flight without refueling. It's likely the tanker crews were being trained to in turn extend the ranges of the fighters and bombers even further in the future."[136]

---

[134] The PLAAF likely did not regularize patrols in the East China Sea, as it did for distant sea training and the South China Sea, because Chinese reporting says that it "has kept regular patrols" since the ADIZ was created in 2013, even if they were not conducted with H-6Ks. See Dong Zhaohui, "China's Air Force Conducts West Pacific Drill, Patrols ADIZ," China Military Online, September 25, 2016; "Multiple Types of Chinese Air Force Aircraft Conduct First Patrol of East China Sea ADIZ" ["中国空军多架战机首巡东海防空识别区"], *Beijing Morning Post* [北京晨报], November 24, 2013; and Zhang Yuwing [张玉清] and Zhao Wei [赵蔚], "Chinese Air Force Spokesperson Shen Jinke Introduces East China Sea Patrols" ["空军新闻发言人申进科介绍东海警巡"], Xinhua, August 7, 2014.

[135] Li Jianwen [李建文] and Yu Hongwei [余泓纬], "Multiple Types of Chinese Air Force Aircraft Fly into Western Pacific to Test Distant Sea Combat Capability" ["空军多型战机赴西太平洋检验远海实战能力"], China Military Online, September 13, 2016.

[136] For one Chinese explanation of the role of tankers in PLAAF long-range operations, see Huang Zijuan [黄子娟], "China's Air Force Tankers Possess Ability to Fly Covertly, Can Fly at Low Sea Levels" ["我军空中加油机具隐蔽出航能力 能海上低空突防"], People's Daily Online, May 18, 2015; see also Kyle Mizokami, "China's Air Forces Are Venturing Further into the Pacific Than Ever Before," *Popular Mechanics*, September 16, 2016.

On September 25, another formation of PLAAF planes flew through the Miyako Strait.[137] However, the exact number of planes that flew through the Strait is disputed, as the Japanese MOD report lists eight aircraft but Shen's announcement claimed 40. A likely explanation is that 40 PLAAF planes were in the formation inside the First Island Chain but only some actually crossed the Miyako Strait. The formation included H-6Ks, Su-30s, tankers, and other unknown aircraft.[138] Although the PLAAF described the flight in similar terms to the November 2015 flight through the Strait—namely, that some bombers and fighters split from the larger group and went into the East China Sea ADIZ to conduct a patrol (警巡)—there appear to be some differences. According to Japanese MOD descriptions of their flight paths, all of the eight plans appear to have taken a generally direct route through the Strait, including the H-6Ks, whereas in November 2015, a group of four H-6Ks broke from the larger formation and flew toward the Japanese main islands before turning away. The PLAAF now appears to consider any H-6K flight that transits the ADIZ on its way through the Miyako Strait as a "patrol" of the ADIZ, possibly for propaganda purposes. The Japanese MOD listed eight total aircraft that transited the Strait: four H-6Ks, one Tu-154, one Y-8, and two planes that it assumed were fighters.[139] The Japanese MOD report suggests that only one Y-8 and one Tu-154 transited the ADIZ. According to *IHS Jane's 360*, a KJ-2000 early warning aircraft was also in the group, judging by China Central Television footage.[140] The planes practiced reconnaissance and early warning, maritime assault on sea-surface targets, and mid-air refueling, among other skills.[141] Before the flight, Shen also said that the PLAAF had made "a qualitative leap of strategic transformation, by intensifying training in realistic scenarios," providing a positive assessment of the training progress.[142]

This was the first time presumed Chinese fighters flew through the Strait, according to the Japanese government.[143] This was also the first time Japan dispatched fighters to intercept the PLAAF planes, and Chinese media noted this change in response. Of note, Chinese planes that

---

[137] For details on the flight, see Japanese Ministry of Defense, "On the Flight of Chinese Aircraft in the East China Sea" ["中国機の東シナ海における飛行について"], September 25, 2016; Ye Pan [叶攀], "40 Aircraft of Multiple Types Go into Western Pacific, China's Air Force Tests Distant Sea Combat Capability" ["40 多架各型战机前出西太平洋 中国空军检验远海实战能力"], Xinhua, September 25, 2016; Slavin, December 13, 2016; and Andrew Tate, "PLAAF Deploys 40 Aircraft to Western Pacific in Show of Strength," *IHS Jane's 360*, September 27, 2016.

[138] Ye Pan, September 25, 2016.

[139] Japanese Ministry of Defense, September 25, 2016.

[140] Tate, September 27, 2016.

[141] Sun Xiaochen and Wang Qingyun, "Air Force in Pacific Drill to Test Far-Offshore Strength," *China Daily*, September 26, 2016.

[142] Tate, September 27, 2016.

[143] Jesse Johnson, "Japan Scrambles Jets as China Warplanes Fly Through Okinawa Strait," *Japan Times*, September 26, 2016; and Qiu Yue [邱越], "Expert: Chinese Air Force Flying into the Distant Seas Forms a System, Initiative Will Expand" ["专家:中国空军赴远海战机形成体系 主动权将扩大"], People's Daily Online, September 28, 2016.

appear to be PLAN Air Force reportedly conducted three flights into Northeast Asia, passing through the Tsushima Strait, in January and August 2016, as well as in January 2017.[144]

Strategic Messaging

Chinese military and media commentary suggested that the flight through the Miyako Strait was intended to pressure Japan and increase the PLAAF's long-distance operational capabilities. Du Wenlong explained that while previous distant sea trainings involved only one type of aircraft and small formations, the late September flight through the Strait included more aircraft and, more important, created a "system."[145] It is unclear why Du would describe previous flights as having only one type of aircraft, as all known PLAAF flights through the First Island Chain except the first two in 2015 included several types of aircraft, according to official statements. Du noted that Japan's decision to mobilize fighters from more than one military base reflected great concern over China's large-scale and long-distance aerial operational capabilities. Du concluded that the capabilities and scope of PLAAF activities in the East China Sea have greatly expanded, and that activities past the First Island Chain will pressure Japan. Contradicting this, Senior Colonel Wang Mingliang said that distant sea training was not targeted at any country but rather was an inevitable trend of the PLAAF protecting China's maritime rights and interests. He added that the service would improve its early warning and reconnaissance capabilities in the distant seas. Suggesting one potential political reason for the flight, at least one Chinese media article tied the flight to Japanese Defense Minister Tomomi Inada's pledge for Japan to patrol the South China Sea with the United States ten days earlier, saying "to protect the South China Sea, the PLAAF will first act (出手) in the East China Sea."[146] Linking the flight back to the PLA's expanding responsibilities away from China's coastline, Zhao Xiaozhuo from the Academy of Military Science said, "The awakening of China's maritime rights consciousness has encouraged the country to expand its activities from coastal areas to far-offshore regions."[147]

The PLAAF has used several different terms to describe similar H-6K activities without clear explanation, likely creating some confusion over the strategic message. The November 2015 flight into the East China Sea ADIZ was referred to as a "distant sea patrol" (远海警巡), while the similar late September 2016 patrol was described as simply a patrol (警巡). Moreover, early

---

[144] For more information on the January 2017 flight, see Japanese Ministry of Defense, "On the Flight of Chinese Aircraft in the East China Sea and Japan Sea" ["中国機の東シナ海及び日本海における飛行について"], January 9, 2017; Kristin Huang, "Japan, South Korea Scramble Fighter Jets After Eight Chinese Warplanes Fly over Korea Strait," *South China Morning Post*, January 10, 2017; and "Chinese Military Planes Flying to Sea of Japan Prompt Scramble," *Asahi Shimbun*, January 10, 2017.

[145] Qiu Yue, September 28, 2016.

[146] Qiu Yue, September 28, 2016; David Brunnstrom, "Japan to Boost South China Sea Role with Training Patrols with U.S.: Minister," Reuters, September 15, 2016; and Jesse Johnson, "Chinese State Media Blasts Japan over South China Sea 'Patrols,' but Experts See No Change in Policy," *Japan Times*, September 18, 2016.

[147] Sun Xiaochen and Wang Qingyun, September 26, 2016.

PLAAF patrols of the ADIZ after it was established in November 2013, without the H-6K, were referred to as "aerial patrols" (空中巡逻), and it appears that the PLAAF began using "patrol" (警巡) in August 2014.[148] The PLAAF described the official H-6K flights into the South China Sea in July and August 2016 as "combat air patrols" or "combat readiness patrols" (战斗巡航).[149] In mid-July 2016, CMC Vice Chairman Fan said that the PLAAF should "organize air and sea patrols" (海空巡逻), but the mid-September 2016 flight through the Bashi Channel practiced "sea surface cruising" (海上巡航), which is likely better translated as "maritime cruising," and authoritative PLAAF books previously discussed "strategic cruising" (战略巡航).[150]

It is unclear what the use of different terms means for PLAAF doctrine and H-6K employment, especially since bombers are generally not considered practical tools for patrolling airspace. However, the appearance of fighters, tankers, and reconnaissance planes in the South China Sea suggests that the PLAAF is seeking to demonstrate that it could enforce an ADIZ or at least establish a presence in the South China Sea, as it has done in the East China Sea. One commentator said that the "combat air patrol" is different from the 2015 ADIZ "patrol" because it is operational and can engage enemy aircraft in combat "anytime during an air patrol."[151]

Flights Circumnavigating Taiwan in 2016

Toward the end of 2016, the PLAAF conducted two flights through the Miyako Strait and Bashi Channel circumnavigating Taiwan. Although these flights have been viewed by some foreign analysts as a signal to Taiwan and the United States of China's commitment to its claims to Taiwan, mainstream Chinese commentary did not explicitly voice this view.[152]

Flight Details

In November 2016, two H-6Ks, one Tu-154, and one Y-8 flew through the Bashi Channel and then back through the Miyako Strait, circumnavigating Taiwan.[153] They were joined by two

---

[148] "Multiple Types . . ." November 24, 2013; and Zhang Yuwing and Zhao Wei, August 7, 2014.

[149] Beauchamp-Mustafaga et al., August 2, 2016.

[150] This "strategic cruising" term has been used to describe Russian long-range strategic bomber flights by Chinese media. See Military Affairs Decoded [军情解码], "The Theory Behind Russian Bombers' Strategic Cruising" ["俄罗斯轰炸机战略巡航幕后玄机"], February 8, 2013; "Fan Changlong: . . . ," July 19, 2016; Shou Xiaosong, 2013, p. 227; Zhu Hui, 2009; Zhang Yuqing and Wu Yi, November 27, 2015; and Zhang Tao, "Chinese Military Aircraft in West Pacific for Combat Simulation Drill," China Military Online, September 13, 2016b.

[151] "Focus Today," China Central Television, July 18, 2016.

[152] Tomlinson, December 9, 2016; and Matthew M. Burke and Chiyomi Sumida, "China Reportedly Responds to Trump's Taiwan Call by Flying Nuclear-Capable Bomber," *Stars and Stripes*, December 12, 2016.

[153] The PLAAF statement does not comment on the types of aircraft involved, and the Japanese MOD report says that only H-6 aircraft were involved. However, the H-6K is only a PLAAF platform, highly suggesting that H-6K

Su-30s that flew out to the Miyako Strait, met up with the group, and flew back through the Strait with the formation without circling Taiwan. U.S. media also reported that J-10s escorted the group at some point in time.[154] China Central Television appears to have also aired footage of KJ-500s and KJ-2000s participating, likely within the First Island Chain.[155] The PLAAF spokesperson appears to have confirmed the flights only following Japanese media reports and described the flights as part of the annual training schedule and not targeted at any country or region. U.S. media cited U.S. officials as saying that the "escort jets were used to collect radar information and conduct other surveillance on American allies such as Japan [and] the Chinese bombers stayed in international airspace."[156] This also marked the first time that mainstream Chinese media picked up Japanese MOD reporting on the flights, implicitly confirming the details provided.[157] Of note, Taiwan was not actually mentioned by the PLAAF announcement, and some Chinese media coverage ignored the Taiwan angle. Taiwan and Japan both scrambled fighters in response.

Less than a month later, in December, two H-6K, one Tu-154, and one Y-8 flew first through the Miyako Strait and then back through the Bashi Channel, again circumnavigating Taiwan but in the opposite direction of the November flight.[158] Two Su-30s flew with the group through the Miyako Strait but then turned around and did not continue around Taiwan through the Channel,

---

were involved. See Japanese Ministry of Defense, "On the Flight of Chinese Aircraft in the East China Sea" ["中国機の東シナ海における飛行について"], November 28, 2016; "Multiple Types of Chinese Air Force Aircraft Fly Through Bashi Channel and Miyako Strait at Same Time" ["中国空军多型战机同时飞越巴士海峡和宫古海峡"], *Observer*, November 27, 2016; and Wei Yunfeng [魏云峰], "Taiwan Media Cry Out over PLA Fighters Training Around Taiwan, Taiwan Military Says Under Control" ["台媒惊呼解放军战机绕台演训 台军方称尽在掌握"], *Global Times*, November 28, 2016.

[154] Tomlinson and Griffin, December 5, 2016.

[155] "Draw a Circle Around Taiwan: China's Air Force Flies into Western Pacific from Two Directions" ["给台湾画个圈 中国空军两个方向前出西太平洋训练"], China.com, November 28, 2016.

[156] Tomlinson and Griffin, December 5, 2016.

[157] "Multiple Types . . . ," November 27, 2016; Kui Jun [隗俊], "English Media: PLA Fighters Flew Around Taiwan, Taiwan Says Closely Monitored" ["英媒:解放军战机环绕台湾飞行 台方称严密监控"], *References News* [参考消息], November 28, 2016.

[158] The PLAAF statement again did not specify H-6K, but Chinese media referred to the plane as an H-6K. See "Air Force Responds, Distant Sea Training, Patrols in East China Sea and Combat Air Patrols in South China Sea to Strengthen Distant Sea Training, Improve Strategic Capabilities" ["空军回应远海训练东海警巡南海战巡:加强远海训练提升战略能力"], China Central Television, December 15, 2016; and Zhang Yuqing and Zhang Mimi, December 15, 2016.

For more information on the December 10, 2016, flight, see Japanese Ministry of Defense, "On the Flight of Chinese Aircraft in the East China Sea" ["中国機の東シナ海における飛行について"], December 10, 2016; Japanese Ministry of Defense, "Emergency Responses in the Third Quarter of FY2016" ["平成２８年度３四半期までの緊急発進実施状況について"], January 20, 2017; and Taiwan Ministry of National Defense, "R.O.C. Military Press Statement Regarding Close Surveillance of People's Liberation Army Far-Offshore Training," December 10, 2016.

heading back through the Strait, according to Japanese MOD reporting.[159] Another group of Su-30s met the formation south of Taiwan to escort them back through the Bashi Channel, according to Taiwanese reporting.[160] The Japanese military scrambled F-15 fighters in response, and Taiwan dispatched a reconnaissance plane to monitor the Chinese planes.[161] Taiwan government reporting suggests that the flight took roughly four hours.[162] Again, PLAAF spokesperson Shen confirmed the flight only following foreign media reporting, but this time the PLAAF released photos on social media several days later providing further evidence of the flight.[163] Although the photographs appeared to show the planes near Jade Mountain and Orchid Island, the Taiwan MND said that they did not pass through its ADIZ, and the spokesperson called the photographs "propaganda," suggesting that they were not taken during the H-6K flights.[164] Although Shen stated in his December 15 announcement that the PLAAF had conducted "distant sea training, a patrol (警巡) in the East China Sea, and a combat air patrol (战巡) in the South China Sea" in recent days, the South China Sea flight appears to have taken place roughly a week before the announcement (December 8).[165] Moreover, Japanese MOD reporting suggests that the ADIZ "patrol" was more like the September 2016 flight than the November 2015 flight—the formation, including H-6Ks, generally took a direct route through the Miyako Strait, and no planes broke formation to fly closer to the main Japanese islands within the ADIZ. Shen had said in November that patrols through the ADIZ would continue.[166]

Strategic Messaging

These flights were touted as a significant marker of progress for PLAAF over-water capabilities. According to H-6K pilot Yang Yong, the flights have evolved through three phases: "single aircraft type flights going out to take a look, formations going out and flying around, and regularizing many aircraft types training beyond the [First] Island Chain."[167] Yang concluded that the two years of flights have made significant contributions for training with the hardware and exercising maritime air control. Du Wenlong asserted that training flights have transitioned from small numbers of aircraft and a few aircraft types to many aircraft of different types, and

---

[159] Japanese Ministry of Defense, December 10, 2016.

[160] Jason Pan, "Chinese Aircraft Fly Around Taiwan," *Taipei Times*, December 11, 2016.

[161] Pan, December 11, 2016.

[162] Taiwan Ministry of National Defense, December 10, 2016.

[163] "PLA Air Force Releases Photos of Seemingly H-6K Next to Taiwan Mountain" ["解放军空军发布疑似轰-6K与台湾中央山脉高峰合影"], *Observer*, December 17, 2016.

[164] "MND Appeals for Calm After Reports of Chinese Aircraft Circling Taiwan," *China Post*, December 19, 2016; and Chen Wei-han, "MND Plays Down China Aircraft Photos," *Taipei Times*, December 19, 2016.

[165] Zhang Yuqing and Zhang Mimi, December 15, 2016.

[166] Li Kaiqiang, November 5, 2016.

[167] Zhang Zhaoying, December 16, 2016.

that the location of patrols and training has moved farther away from the coastline.[168] He noted that the December 2016 flight had "almost 10 aircraft, including fighters, bombers, early warning aircraft and tankers," which demonstrated that the "larger air combat system is already formed." Taken together with the fact that training activities are becoming more frequent and attack capabilities are improving, this means that the PLAAF's "combat capabilities are increasingly close to the strategic goals." Echoing this, PLAAF expert Fu Qianshao said, "compared with the first patrols by the Air Force, recent ones featured more aircraft, longer ranges and participation from multiple theater commands."[169] Rear Admiral Yin Zhuo added that flying out and return through different routes was very challenging but showed that the PLAAF's combat capabilities had improved and that the PLAAF was now prepared to fight in the East and South China Seas at the same time.[170] Looking forward, Wang Ya'nan, editor of the PLAAF's *Aerospace Knowledge* magazine, projected that "the Air Force will send early warning, aerial refueling and electronic warfare planes to take part in future long-range operations."[171]

Maritime Strike Training

Beyond these long-range training flights for the H-6K and other aircraft, the PLAAF has also expanded maritime strike training in recent years for all coastal regions. For example, a Guangzhou MR H-6K unit practiced distant sea missile launches in September 2015.[172] The unit flew "several thousand" kilometers from southern China over water to conduct "low-level flights to interdict 'enemy' sea defenses [and strike] long-distance targets" with multiple missiles. State-run media said that this "signifies that H-6K units already have the capability to launch precision strikes anywhere in Chinese territory under any weather conditions."[173]

The Nanjing MR conducted several over-water exercises in 2015. One Nanjing unit claimed that "most pilots' above-the-sea ultra-low-altitude continuous flight time exceeded the higher range of time specified by the outline" and in exercises had practiced "accurately [firing] guided

[168] Zhang Zhaoying, December 16, 2016.

[169] Zhao Lei, "Air Force's Patrols to Continue over East, South China Seas," *China Daily*, December 16, 2016.

[170] Qiu Yue [邱越], "Expert: Air Force Already Has Ability to Respond to Operational Needs in East and South China Seas at Same Time" ["专家:空军已有能力同时应对东海南海作战需求"], People's Daily Online, November 29, 2016.

[171] Zhao Lei, December 16, 2016.

[172] Although most of the articles covering this training were from October, the flight appears to have taken place in late September. See Yu Hongwei [余泓纬] and Xu Sheng [徐生], "Guangzhou Air Force H-6K Have Capability to Launch Precision Strikes Anywhere in Chinese Territory Under Any Weather Conditions" ["广空轰-6K 部队具备全疆域全天候远程精确打击能力"], China News Online, September 30, 2015; and "Guangzhou Air Force H-6K Have Capability to Launch Precision Strikes Anywhere in Chinese Territory Under Any Weather Conditions" ["广空轰-6K 部队具备全疆域全天候远程精确打击能力"], People's Daily Online, October 13, 2015.

[173] Yu Hongwei and Xu Sheng, September 30, 2015.

missiles when launching precision strikes [for] hitting mobile targets at sea."[174] In December, a Chengdu MR unit participated in maritime training over 1,000 km from its base, but no other details were provided.[175]

In May 2015, a Shenyang MR unit conducted "combat-realistic" over-water training that "led to a five percentage improvement…in covert penetration, search and identification, and distant sea attack, and a ten percent improvement in live sea target hit rates."[176] This was part of the unit's 2015 training emphasis on "building a routine ground and sea defense penetration and attack training system." The unit also practiced for challenging weather that could "disorient" a pilot and pilots' ability to perform "under complex electromagnetic environment," the PLA's description of electronic warfare (EW). Another training exercise by the same unit, apparently in June, practiced "radio silence to [conduct] low and very low-altitude covert penetration," avoiding enemy anti-aircraft weapons, and "[conduct] point-target attacks on numerous targets in unfamiliar waters." In December, another unit trained for "low altitude penetration evasion of anti-aircraft weapons" in radio silence.[177]

The PLAAF continued maritime training across theater commands in 2016. Perhaps the most ambitious training exercise was in February 2016, when a group of 20 bombers, from either a single or a combination of TCs, conducted long-range penetration and assault training through four theater commands and covered over 10,000 km in 35 hours.[178] The flights included "simulated bombings and missile attacks against multiple types of targets on land and at sea from daytime to nighttime." The training was partly a test of a new task planning system, which "can be used for battlefield posture analysis and calculation, for precisely planning the use of weapons and ammunition, [and] for designing penetration and assault routes," as well as to "provide whole-process information support for reconnaissance, control, striking, and effects assessment." Reflecting the training's effectiveness, the front page *PLA Daily* article stated, "In only two months after the system was put into use, a regiment of the division successfully achieved the

---

[174] "China Launches Intensive Training for Several Types of Fighters: Su-30 Flying Leopard Eye-Catching" ["中国多型战机密集升空训练苏 30 飞豹抢眼"], China News Online, March 4, 2015; and Liao Qirong [廖启荣], "Build the Elite Force Along the Course Specified by the Rules: Investigation and Consideration on the Practice of an Aviation Division of the Nanjing Military Region Air Force in Raising the Training Rules Execution Power in the Course of Strictly Conducting Training in Accordance with the Outline" ["雄帅之路依法而行:南空航空兵某师按纲施训从严治训提升训练法规执行力的调查与思考"], *Kongjun Bao*, February 10, 2015.

[175] Xu Bianchun and Yang Chaoli, December 17, 2015.

[176] Xue Xiaoling [薛小岭] and Chen Yi [陈益], "Combat-Realism Becomes Routine: Shenyang MRAF Aviation Brigade Enhances Training Quality of Over-Land and At-Sea Defense Penetration and Attacks" ["实战化进入常态化: 沈空航空兵某旅提升陆地海上突防突击训练质效"], *Kongjun Bao*, June 9, 2015, p. 2.

[177] Xue Xiaoling [薛小岭], "Shenyang PLAAF Unit Conducts Flight Training" ["沈空航空兵某旅进行飞行训练"], China Military Online, December 15, 2015.

[178] Wei Bing [魏兵] and Cao Chuanbiao [曹传彪], "Aviation Division of the Air Force Applies a Task Planning System to Realistic Combat Training According to Actual Missions, Precisely Linking Air Battle Training to the Actual Battlefield" ["空军航空兵某师运用任务规划系统推进实战化训练"], *PLA Daily*, February 18, 2016, p. 1.

result of delivering the most accurate attacks on targets at the longest range by breaking through multiple layers of interception under an extreme weather condition, as they applied the task planning system to adeptly plan the flight route."

In April 2016, the Western TC participated in a foreign exercise, very likely "Shaheen V" with Pakistan, focused on "air defense and counteroffensive operations," which included an over-water component.[179] In July 2016, a Western TC aviation brigade conducted an "aerial surveillance patrol" and air combat training with an over-water component.[180]

In May 2016, the Northern TC for the first time conducted a maritime penetration and assault exercise with all-female pilots, known as the Flying Leopards.[181] The exercise included strikes on maritime targets:

> During the exercise, to dodge the "enemy" air defense fire, female pilot Wai Xue adopted the form of ultra-low-altitude maneuvering to successfully evade the opponent's radar search. Being supported by comrade-in-arms Gao Fuyu's crew who staged a "feint," she climbed up swiftly to launch attacks, moved overhead the target sea area, and sank the maritime "target" in one strike.

In July 2016, other units from the Northern TC conducted another maritime penetration and assault exercise, this time with dissimilar aircraft, which included strikes on maritime targets.[182]

In July 2016, the Eastern TC conducted airborne refueling training over water, which was followed by some pilots conducting air combat training while others "flew their aircraft in formation to a certain sea area to carry out long-duration flight and long-distance combat warning patrol tasks at sea."[183] According to one deputy squadron commander, "developing

---

[179] Pete Sweeney, "Air Forces of China, Pakistan Launch Joint Training Exercise," Reuters, April 9, 2016; and Cao Chuanbiao [曹传彪] and Zhang Li [张力], "How Many Things Need to Be Coordinated in the Relationship Between Combat Operations and Force Building: The Practice and Thought of the Air Force of the Western Theater Command in Adapting to the New Leadership and Command System and Pushing Forward Realistic Military Training" ["战与建,有多少关系需理顺: 西部战区空军适应新的领导指挥体制推进实战化训练的实践与思考"], *PLA Daily*, October 31, 2016, p. 7.

[180] Zhou Shiyu [周士瑜], "From 'Training for Fighting' to 'Training Being Fighting': Sidelights of Realistic Combat Drills Between Opposing Forces Conducted by an Aviation Brigade of the Western Theater Command's Air Force" ["从'训为战'到'训即战': 西部战区空军航空兵某旅组织实战化对抗训练侧纪"], *Kongjun Bao*, August 3, 2016, p. 3.

[181] Li Weisheng [李为胜] and Hao Maojin [郝茂金], "Female Flying Warriors Train over the Territorial Sea: Air Force Aviation Regiment of the Northern Theater Command Carries Out a Maritime Penetration and Assault Exercise" ["飞天女将试剑海疆: 北部战区空军航空兵某团执行海上突防突击演练侧记"], *Kongjun Bao*, May 27, 2016, p. 1.

[182] Yan Xinyu [严新宇] and Yang Pan [杨盼], "Northern Theater Command Air Force Aviation Division Coordinates with Multiple Types of Military Aircraft in Defense Penetration and Assault Drills" ["北部战区空军航空兵某师: 协同多形战机演习突防突击"], *Kongjun Bao*, June 6, 2016, p. 1.

[183] Yue Mingliang [岳明亮], Su Zhibo [苏志波], and Ding Yibo [丁义波], "Practical Training: 'Refueling +;' Account of Eastern Theater Command Air Force Air Brigade Organizing Key-Element Integrated Training with Tactical Backdrop" ["实训: '加油+': 东部战区空军航空兵某旅组织战术背景下要素融合训练侧记"], *Kongjun Bao*, August 4, 2016, p. 2.

skills to improve aerial refueling capabilities allows us to shed the disadvantages of short distances, engage in air combat at even further distances, have even longer warning patrols, and extend combat power even further to distant seas."

In August 2016, the Central TC conducted low-altitude distant sea maritime strike training that represented breakthroughs for the number of aircraft and length of the flight.[184] According to media reporting, "after reaching the [unidentified] island for airdrops, they successfully accomplished the mock airdrop action, then climbed up for the return journey, and carried out accurate airdrops in kind over the target zone on land," and one pilot noted that the most challenging aspect "was the low-altitude and ultra-low-altitude flight at distant sea for the first time."

The Southern TC, the vanguard of the PLAAF's maritime efforts, now includes all aviation units in maritime training. An October 2016 article described a maritime strike exercise in which "several airplanes of three generations, after in-flight refueling, rapidly shaped a combat formation, skimmed the rolling sea surface over a long distance at a low altitude, and then swooped directly toward the 'enemy' targets at sea."[185] Every year, the Southern TC has its pilots "practice low-altitude or ultra-low-altitude penetration in distant sea areas with tactical collaboration of multiple types of aircraft against unknown targets at sea and in the air against complex electromagnetic jamming according to the requirements of actual operations." Clearly maritime strike training extends beyond the H-6K.

Potential PLAAF Long-Distance Strike Targets

As the recent training flights have demonstrated, the H-6K's extended range compared with that of its predecessors enables the PLAAF to fly farther from China, expanding the list of potential targets for the PLA. One specific application of this extended range, which is often unreported in Chinese official statements and state-run media, is the H-6k's strike capabilities against Guam. In its 2010 report on the Chinese military, DoD said, "the PLAAF is developing long-range versions of the B-6/BADGER bomber [now known as the H-6] that, when equipped with a long-range land-attack cruise missile, will enable strikes as far as the second island chain."[186] The 2016 report updated this to note that "the PLAAF also employs the new, fully

---

[184] Zhou Jianhua [周建华] and Tian Yu [田宇], "Air Force Division of the Central Theater Command Comprehensively Forges Power Projection Capabilities by Challenging Maritime Training Extremes" ["中部战区空军某师全方位锤炼投送能力: 海上训练摸索挑战极限"], *PLA Daily*, August 17, 2016, p. 12.

[185] Fan Yishu [范以书], Zeng Ke [曾科], and Li Jianwen [李建文], "The Southern Theater Command Air Force Enhances Maritime Combat Capabilities According to the Requirements of Missions and Tasks by Extending the Sphere of Operations from Near Seas to Distant Seas" ["南部战区空军着眼使命任务提升海上作战能力: 跨越: 从近海到远海"], *PLA Daily*, October 13, 2016, p. 4.

[186] Office of the Secretary of Defense, *Annual Report to Congress: Military and Security Developments Involving the People's Republic of China 2010,* Arlington, Va.: U.S. Department of Defense, 2010, pp. 33–34.

redesigned H-6K variant with new turbofan engines for extended range and the capability to carry six LACMs [which] gives the PLA a long-range standoff offensive air capability with precision-guided munitions capable of striking Guam."[187] Citing foreign media in 2015, the state-run China Military Online said that the "H-6K bomber has a combat radius of 3,500 km. Plus the range of KD-20 LACM, it can cover the U.S. military bases in Northeast Asia and Guam," a rare mention.[188] Moreover, the 2013 edition of the *Science of Military Strategy* states a desire to be able to strike the Second Island Chain.[189] According to a U.S. congressional report, "a senior Japanese military official" asserted that the November 2015 flight "was likely a simulated attack on Guam."[190]

However, the H-6Ks face some challenges for a potential attack against Guam. According to one report, "Due to range (fuel) limitations these bombers would not be able to fly low radar-evading flight paths on their way out past the first island chain. Only by flying high in less dense air could they hope to get close enough to launch their LACMs on Guam."[191] Moreover, they "would be easy targets for American, Japanese, and Taiwanese air defenders long before they could get within range of Guam," as PLAAF fighters cannot fly that far to escort them.[192]

The PLAAF may also be training for strikes against India. It should be noted that the PLAAF's long-distance strike exercises have not been exclusively over water. In December 2015, a Guangzhou MR H-6K bomber unit flew to a "plateau target range" and practiced low-altitude penetration and long-range air-to-ground strikes.[193] This was one of the units that overflew the Bashi Channel out to the First Island Chain in March and August 2015, highlighting the PLAAF's interest in developing capabilities not just for a conflict over the Pacific but also in South or Central Asia, most likely against India.

Last, Australian experts have voiced concern over the H-6K threat now that the Chinese-built features in the South China Sea reportedly host runways long enough for the H-6K, along with

---

[187] Office of the Secretary of Defense, April 26, 2016, p. 31.

[188] Zhang Tao, February 25, 2015.

[189] The *Science of Military Strategy* says, "The effective operational radius should reach 3000 km beyond the borders, so that platform radius or platform radius plus firepower radius reaches the Second Island Chain, to provide capability support for expanding the air defense areas and for repelling from as far as possible the enemy's air and space attack. . . . Precision firepower strike should be able to . . . implement the effective destruction of fixed targets on land and moving targets at sea, to effectively cover the First and Second Island Chains in the near future and gradually expand to cover part of the Indian and Western Pacific Oceans in the mid- to long term." See Shou Xiaosong, 2013.

[190] Wilson, May 10, 2016, p. 9.

[191] Ian Easton, "China's Evolving Reconnaissance-Strike Capabilities: Implications for the U.S.-Japan Alliance," Arlington, Va.: Project 2049 Institute, February 2014, p. 26.

[192] Easton, February 2014; and Wilson, May 10, 2016, p. 9.

[193] Xu Sheng [徐生], "War Eagle Tough Training for Hard Power Signifies Blue Sky" ["战鹰振翅苦练硬功显威蓝天"], China Military Online, January 3, 2016.

aircraft hangars to shelter them.[194] Australian media reported that Malcolm Davis of the Australian Strategic Policy Institute expressed concern over these developments, and Stephan Freuhling, a professor at Australian National University, said that "China's trying to send a signal that, if Australia gets involved directly or indirectly in joint patrols in the South China Sea, Australia shouldn't assume that its distance protects it" and future patrols off Western Australia "would be quite useful to send that signal.[195]

---

[194] David Wroe, "Chinese Bomber Planes from South China Sea and Future Missiles Could Threaten Australia," *Sydney Morning Herald*, September 3, 2016; and "Build It and They Will Come," Washington, D.C.: Center for Strategic and International Studies, August 1, 2016.

[195] Wroe, September 3, 2016.

5. Joint Service Maritime Training

One central challenge in the PLAAF's new emphasis on maritime operations is its ability to coordinate with the PLAN. According to one U.S. government analyst, "true joint interoperability remains largely a work in progress for the PLA" in general, and "the most glaring [training shortcoming] is the limited joint training the PLAAF conducts."[196] PLAAF-PLAN joint operations in a future conflict "would likely consist of the PLAAF providing coincident air cover for PLAN units, and their actions would be deconflicted with naval aviation by location and timing," and "due to a lack of practiced interoperability, their efficiency will decline as they have to react to a dynamic environment and rapidly changing battlespace conditions."[197] Addressing these issues in the maritime domain by citing President Xi's call to be able to fight and win wars by 2020, the *PLA Daily* called for the "building of a maritime combat operations system."[198] This included

> [constructing] an information-firepower closed loop integrating reconnaissance, control, attack, and evaluation, fuse, connect, and integrate maritime attack measures which are scattered among the Army, the Navy, and the Air Force, and form up an inshore waters comprehensive management and control system which can monitor and control the situation at sea at all times and bring crises at sea under control promptly and which can respond quickly and integrate joint action.

Although this type of activity has become more routine in recent years in a number of mission areas, the PLAAF's experimentation and training only recently has involved more robust operations with the PLAN. This effort to improve its joint operations capability with the PLA Navy Air Force is most notably in Sharp Sword-2015.

Sharp Sword-2015

The Sharp Sword-2015 exercise was held in November 2015 and involved "fighter jets, fighter bombers, bombers, airborne early warning and control systems, reconnaissance planes, and helicopters."[199] Although no specific location was provided, the exercise was organized by

---

[196] Fuell, January 30, 2014.

[197] Fuell, January 30, 2014.

[198] Gao Feng [高峰] and Zuo Dengyun [左登云], "Exert Force with Precision, Speed Up the Building of a Maritime Combat Operations System" ["精准发力, 加快海上作战体系建设"], *PLA Daily*, July 31, 2016, p. 7.

[199] Qiu Yue [邱越], "Expert: Joint Navy and Air Force Exercise Boosts Strength to Seize Air Supremacy and Gives Play to the Navy's Ability to Attack; Chinese Air Force's H-6K Bombers Take Part in 'Sharp Sword-2015 [lijian-2015]' Real-Troop Exercise" ["专家:海空军联演助力制空权争夺 发挥海军打击能力"], People's Daily Online, November 4, 2015.

the Guangzhou MR, and photos in state-run media depict fighters over land features in an ocean, confirming that over-water training was part of the exercise.[200] We identified these as Discovery Reef and Passu Keah in the Paracels in the South China Sea. PLAAF units from the Guangzhou MR and PLAN and PLAN Air Force units from the South Sea Fleet participated, involving J-10, J-11B, and H-6K aircraft, among others, and the focus was on system-of-systems confrontations and modern information warfare.[201] Specifically, this included "surveillance and early warning, command and control, fire attack, information warfare, and comprehensive defense"; "acquiring, transmitting, and distribution of information to precision strikes, impact assessments, and second-wave attacks"; and "air-to-surface attack targeting at vessels on the sea or in the ports."[202] The exercise was intended to "allow various units to experience combat training in unfamiliar environments and prepare them for future wars."[203] It does not appear that a Sharp Sword exercise was held in 2016.

One military expert drew three highlights from the exercise: "three-dimensional offense and defense on the high seas"; "the air combat system has been improved to meet the needs of information warfare, especially the practical needs of maritime military struggle"; and last, H-6K "[cooperation] and [coordination] with many other types of tactical aircraft and support aircraft from the Navy and the Air Force in comprehensive trainings on the high seas."[204] This shows "the capability of the PLA Air Force in strategic strike, tactical support, cooperation and coordination on the high seas." Another military commentator framed the exercises in terms of air superiority as the key for winning the naval battle, reflecting authoritative PLAAF writings: "With the support of the Air Force, the Navy will be able to seize air supremacy over a large area more easily in times of war. Seizing air supremacy is the first thing to do in any war. Only when one commands air supremacy, can its aircraft carrier, large-sized destroyers, and submarines give play to their tremendous capabilities to attack seaborne and ground targets."[205] Another military expert added that with its "long-range and accurate striking capacity, [the] H-6K is capable of destroying rival's strategic targets on the land, thereby creating a favorable condition for the navy."[206]

---

[200] Chinese military commentator Yin Zhuo also said that the training "mainly took place at sea." See Qiu Yue, November 4, 2015; Ma Junqing [马俊卿], "Several Types of Air Force and Navy Fighters Participate in 'Sharp Sword–2015' Military Exercises" ["我空军海军多型战机参加'利剑－2015'军实兵演习"], Xinhua, November 3, 2015.

[201] Zhang Tao, November 4, 2015.

[202] Qiu Yue, November 4, 2015; and "Focus Today," China Central Teleision, November 3, 2015.

[203] Wang Baofei [汪保飞] and Yang Qingyue [杨清悦], "Guard Against 'Loss of Acclimatization' of Combat Mentality" ["谨防作战思维'水土不服'"], *PLA Daily*, December 6, 2015.

[204] Zhang Tao, November 4, 2015.

[205] Qiu Yue, November 4, 2015.

[206] "Focus Today," November 3, 2015.

PLAAF Bomber Coordination with PLAN Ships

Suggesting some progress in PLAAF-PLAN joint operation efforts, PLAAF bombers practiced coordinating with PLAN ships sometime in the second half of 2015. A Guangzhou MR PLAAF division "organized two rounds of training along with the Army and the Navy, and changed their understanding of joint training. One time, the division researched offensive tactics along with a certain naval flotilla. With target information provided by the naval command ship, bombers over the sea had 'sharp ears and keen eyes.'"[207] Another article likely discussing the same training exercise provided further detail:

> In one session, together with a naval unit, they jointly practiced tactics of assaulting maritime targets. With target information provided by the naval force, the Air Force bombers were able to successfully carry out the assault mission. In another session, together with an Army brigade, they jointly practiced the obstacle clearing mission in a landing operation, and settled the difficult issue of causing secondary obstacles in the course of clearing up obstacles by means of joint fire strikes.[208]

Expounding on the benefits of joint operations for long-distance bomber operations over water, an article in *Kongjun Bao* remarked, "Now, if bomber units keep their eyes looking outward, stay focused on combat requirements, leave conventional training behind, fly far out over the sea, move in the direction of jointness, and explore new ways of tri-service merged training," they will be able to "carry out well and continually expand their missions and tasks."[209] Another article added that, "compared with battles on land, maritime air operations in the future will be joint operations under informatized conditions. Such combination of combat power is not the simple aggregation of multiple aircraft types and multiple arms of the Air Force and the Navy, but in-depth integration of their operation notions, power systems, and action patterns."[210]

By June 2016, joint training for the Eastern TC PLAAF had been "normalized" and "the air zone for training was expanded from airspace above land to airspace above the sea."[211] A May 2016 online joint training with the Army included PLAN hovercraft and PLAAF bombers, and when the Army red force leader requested fire support from two PLAAF bombers against enemy artillery, "the request was turned down by the exercise direction group. The reason was that the

---

207 Mou Xingguang [牟星光] and Chi Yuguang [迟玉光], "An Aviation Division of the Eastern Theater Air Force Breaks Free of Constrained Thinking, Sets Out on a New Path of Joint Training" ["训练需求表"向实战深度融合:东部战区空军航空兵某师打破思维局限迈开联合训练新路"], *Kongjun Bao*, April 8, 2016, p. 1.

208 Xu Sheng [徐生] and Chi Yuguang [迟玉光], "Aviation Division of the Air Force of the Eastern Theater Command Normalizes Joint Training" ["东部战区空军航空兵某师:联合训练走向常态化"], *PLA Daily*, June 6, 2016, p. 1.

209 Mou Xingguang and Chi Yuguang, April 8, 2016.

210 Fan Yishu, Zeng Ke, and Li Jianwen, October 13, 2016.

211 Xu Sheng and Chi Yuguang, June 6, 2016.

red force did not clearly provide the timing and directions for Air Force bombers' entry into and withdrawal from the operation zone, and this made it impossible to carry out air and ground collaboration."[212] This suggests that as joint operations training increases, the standards for coordination among services will improve.

Other Domestic Joint Training

Another domestic joint exercise took place in September 2015 in the East China Sea.[213] PLAAF units from the Nanjing MR, featuring Golden Helmet winners and "domestically produced third-generation aircraft," and PLAN Air Force units from the East Sea Fleet using "imported third-generation aircraft," practiced air confrontations.[214] This was the first time the PLAAF and PLAN Air Force conducted joint free air combat exercises using different aircraft and followed the 2014 exercises, which were the first for PLAAF-PLAN Air Force free air combat in general. Specifically, they trained for combat under EW attacks, using different runways, "over-the-horizon attack," and "medium-range interception" and included a PLAAF-organized "dogfight skill competition."[215] This was linked to "ongoing air battle drills between the Air Force and Navy as opposing forces."[216] Similar PLAAF-PLAN Air Force long-distance strike training appears to have taken place in April 2016, and PLAN Air Force pilots participated in Golden Helmet 2015, held over water.[217]

---

[212] Chen Dianhong [陈典宏] and He Weili [何伟理], "Shape a 'Joint Battlefield,' Foster the 'Strongest Brain': Army Divisions and Brigades of the PLA Southern Theater Command Conduct Authentic and Complex Online Command Confrontation Drills" ["构'联合战场' 练'最强大脑': 南部战区陆军合成师旅网上指挥对抗演练逼真复杂"], *PLA Daily*, May 9, 2016, p. 4.

[213] Li Kaiqiang [李开强] and Wu Yi [伍佚], "'Golden Helmet' Kills 'Combat Elite'" ["'金头盔'PK'空战精英':海空军航空兵首次异型机自由空战对抗演练侧记"], *PLA Daily*, October 12, 2015; and Zhang Heng [张恒], "Confrontation Between Different Models of Aircraft Is Carried Out in a 'Back-to-Back' Way: The Joint Training of the Aviation Forces of the Navy and the Air Force Is Expanding in Depth" ["异型机对抗背靠背进行:海空军航空兵联合训练向深度拓展"], *PLA Daily*, September 20, 2015.

[214] The PLA's "third-generation" aircraft are considered by Western standards to be fourth-generation aircraft. See National Air and Space Intelligence Center, 2010, p. 71; Li Kaiqiang and Wu Yi, October 12, 2015; and Zhang Heng, September 20, 2015.

[215] Li Jianwen [李建文] and Lu Hui [卢辉], "Aviation Division of the Nanjing Military Region Air Force Uses the Military Parade Spirit to Arouse the Personnel's Combat Readiness Enthusiasm, Painstakingly Practices Skills in Flight Training, Raises Actual Combat Capability" ["苦练精飞提高实战能力"], *PLA Daily*, September 19, 2015; and Zhang Heng, September 20, 2015.

[216] Li Jianwen and Lu Hui, September 19, 2015.

[217] "Air Force and Navy Air Force," China Central Television, April 21, 2016; Chase, Allen, and Purser, 2016, pp. 12–13.

Joint Sea-2015 II

The two services also cooperated in Joint Sea-2015 II for nine days in August 2015, China's largest naval exercise with Russia since the exercises began in 2012.[218] The PLAAF sent two J-10s, two JH-7s, and one KJ-200 early warning aircraft to participate in "joint air-defense" and "joint [amphibious] landing."[219] The air-defense part included the KJ-200 providing "command and guidance" to the landing forces while the J-10s and JH-7s intercepted and drove away the opposition blue forces, a Russian Navy Pacific Fleet IL-38, and TU-142M aircraft.[220] The amphibious part was centered on air-to-ground support against the shore, with the J-10s providing strike cover to the landing troops while the JH-7s bombed the landing approaches, supported by strikes from four Russian Su-25s. According to one Chinese military expert, this was the first time the PLAAF and PLAN conducted a joint exercise away from China, and it was PLAAF's first time participating in Joint Sea with Russia. Joint service cooperation for future warfare is critical because "relying only on the PLA Navy, [China] will be unable to rule the seas and win a war."[221]

Chinese media coverage emphasized China-Russia mutual trust and cooperation, as well as some linkage of the amphibious landing training to China's territorial disputes. One state-run article said, "Russia and China both have their own disputed islands, and these disputes are a factor in uncertainty in the peace and stability of the Western Pacific region. China and Russia's 'Joint Sea' maritime exercise serves the need of safeguarding regional security and stability, and of course these things will be considered and reflected in the training topics and location."[222] Although coverage was careful to point out that "the drill did not set up [an] imaginary enemy, and was not against any specific target," this landing exercise would be applicable to future

---

[218] Sam LaGrone, "China, Russia Land 400 Marines in First Joint Pacific Amphibious Exercise," *USNI*, August 26, 2015; Liu Guangzhi [刘广智], Tian Xuhui [田旭辉], and Liu Daquan [刘大全], "China's Air Force Mobilizes Three Types and Five Planes to Participate in China-Russia Maritime Exercise" ["中国空军出动 3 型 5 架战机参加中俄海空联合军演"], China Military Online, August 24, 2015; Ren Xu [任旭] and Xu Yeqing [徐叶青], "China-Russia Militaries Successfully Conduct Joint Landing Operation in Exercise" ["中俄两军联合立体登陆演习成功举行"], China Navy Online [中国海军网], August 26, 2015; Zhang Zhongkai [张钟凯] and Wu Dengfeng [吴登峰], "Military Expert Explains Highlights of China-Russia 'Joint Sea-2015 (II)' Exercise" ["军事专家详解中俄'海上联合-2015(II)'演习亮点"], Xinhua, August 24, 2015; Li Dapeng [李大鹏], "What Breakthroughs Does Joint Sea-2015 China-Russia Exercise Reveal?" ["海上联合-2015"中俄联演实现了哪些新突破"], China Youth Daily Online [中国青年网], October 9, 2015; Chen Haifeng [陈海峰], "Joint Sea 2015 (II) Military Exercise Starts Today, China and Russia Exercise in Sea of Japan" ["海上联合-2015(II)军演今起举行 中俄将在日本海联演"], China News Online, August 20, 2015; and Yu Hongwei [余泓纬], "Guangzhou H-6K Possesses Ability to Conduct Long-Distance Precision Strikes Anywhere in Chinese Territory" ["广空轰-6K 部队具备全疆域全天候远程精确打击能力"], China Military Online, October 12, 2015.

[219] Liu Guangzhi, Tian Xuhui, and Liu Daquan, August 24, 2015.

[220] Li Dapeng, October 9, 2015.

[221] Li Dapeng, October 9, 2015.

[222] Li Dapeng, October 9, 2015.

scenarios on Taiwan, the Senkaku/Diaoyu Islands or the South China Sea, as the exercise was intended to "show the conviction and resolve of the two countries to jointly deal with maritime safety threats, and safeguard world maritime security hand in hand."[223]

While the PLAAF and PLAN Air Force may conduct more joint training as part of foreign exercises in the future, these will likely be limited to Russia in the near future for any significant demonstration of capabilities. While there was no coverage of specific skills or tactics the PLAAF could adopt from the Russian military, Zhang Zhaozhong, a PLA commentator, said, "we can have exchanges of data that go down to deep levels during operations, with reciprocal command during operations, these sorts of things."[224] He further noted that such deep cooperation was "impossible" with the United States and the West, as "they would not feel at ease in exchanging data with us, what their sonar sees, what their radar sees. Will they tell us what their radar sees, how far it can see, at what distance can they clearly see targets? It is not possible that they would transmit that data to us."

It does not appear that the PLAAF participated in Joint Sea 2016, which was hosted by the PLAN South Sea Fleet and conducted in the South China Sea in September 2016.[225]

PLAAF-PLAN Air Force Maritime Strike Mission Overlap

Although the PLAAF and PLAN Air Force both appear to have maritime strike missions, there appears to be little effort at higher levels of PLA leadership to disentangle these overlapping responsibilities.[226] The 2016 DoD report on the PLA said that these "inter-service exercises and real-world operations…suggest that China is seeking to integrate future air operations. Integrating air power could allow China to enhance flexibility of strike and support aircraft in East China Sea, South China Sea, or Taiwan contingencies."[227] The joint exercises are intended to demonstrate the services' progress toward these goals.

---

[223] Yu Jinhu [余金虎], "Ten Series of Military Exercises Have Promoted the Innovation and Upgrading of PLA's Combat Effectiveness" ["十大系列军演推动解放军战斗力创新升级"], China Military Online, March 12, 2016.

[224] "Defense Review Week," China Central Television, September 11, 2016.

[225] Zhang Kejin [张科进] and Chen Guoquan [陈国全], "Sino-Russian 'Joint Sea-2016' Exercise Kick Off Today" ["中俄'海上联合-2016'演习今起举行"], *PLA Daily*, September 12, 2016; Zhang Tao, "Chinese and Russian Troops Ready for Joint Sea 2016 Drill," China Military Online, September 13, 2016a; "China, Russia Start Joint Navy Drill in South China Sea," Xinhua, September 13, 2016; and Zhang Junshe [张军社], "China-Russia Naval Exercise Reflects Strategic Mutual Trust Between the Two Countries" ["中俄海上军演展现两国战略互信"], *People's Daily Overseas Edition* [人民日报海外版], September 13, 2016, p. 1.

[226] For background on a potential anti-ship mission for the PLAAF, see Cliff et al., 2011, p. 183.

[227] Office of the Secretary of Defense, April 26, 2016, p. 68.

6. Conclusion

Although the PLAAF's leaders had recognized the importance of developing into a strategic air force since the early 2000s, if not sooner, President Xi's call for improved PLA capabilities to safeguard China's maritime security interests accelerated the PLAAF's transition from homeland defense to regional maritime power projection. Recent PLAAF training reflects this shift. Its emphasis on long-distance patrols and long-distance strike missions out to the First Island Chain represents an unprecedented body of activity for the PLAAF. This activity also serves as an initial glimpse into what many PRC military and political leaders would like to see the PLAAF become—a force capable of projecting power over both the land and maritime domains. China's expanding interests and the need to develop actual military capabilities drive these new imperatives. The PLAAF will have to expand its horizons if for no other reason than its need to maintain relevance in light of an evolving international security situation in which senior Chinese leaders perceive a steady increase in threat to PRC maritime claims. Accordingly, PRC decisionmakers will look to the PLAAF to enforce territorial claims, support strategic conventional deterrence, and, in the case of war, prosecute maritime strikes.

The recent long-distance flights into the Pacific and South China Sea mark a training progression toward increasingly frequent and complex flights and suggest that the PLAAF is transitioning from the experimental phase to regularizing these long-range power-projection activities as PLAAF capabilities mature. This report has highlighted a series of critical exercises and experimentation events designed to clarify and solve key operational questions. The process designed to tackle these problems is thorough, deliberate, and time-consuming. The solutions and subsequent operational concepts will not emerge overnight; however, eventually, these new ideas will crystalize and be regularized among PLAAF units and aircrews.

Currently we are unable to identify the specific components of these operational concepts. The PLAAF routinely has discussed its new concepts in system-of-systems terminology, implying the need for joint operations. Recent PLAAF-PLAN joint exercises and Sharp Sword-2015 suggest that any future concept will be embedded in a joint framework. Furthermore, the conclusion of a Sharp Sword exercise also suggests that additional exercises will follow and dive into the specific operational research problems identified during this key exercise series. We currently lack specifics on what these may be, but previous iterations have highlighted a robust follow-on effort to build on these initial findings.

The PLAAF's inability to perform these over-water missions has detracted from its status as a strategic air force. Although there had been some debate in recent years about what this concept really meant, direction from Xi and other top PLA leaders has provided unprecedented clarity in at least one key area—maritime strike. The PLAAF's success or failure at implementing these changes will bear directly on its future strategic relevance.

References

"Air Force and Navy Air Force," China Central Television, April 21, 2016. As of July 17, 2017: http://tv.cctv.com/2016/04/21/VIDEb55Uo89fLcPw1HFe9WuL160421.shtml

"Air Force Responds, Distant Sea Training, Patrols in East China Sea and Combat Air Patrols in South China Sea to Strengthen Distant Sea Training, Improve Strategic Capabilities" ["空军回应远海训练东海警巡南海战巡:加强远海训练提升战略能力"], China Central Television, December 15, 2016. As of July 17, 2017: http://news.cctv.com/2016/12/15/ARTI2Xjfn6wYAqW68lSKrCr6161215.shtml

Allen, Kenneth W., Dennis J. Blasko, and John F. Corbett, Jr., "The PLA's New Organizational Structure: What Is Known, Unknown and Speculation (Part 1)," *China Brief*, February 4, 2016. As of July 17, 2017: https://jamestown.org/wp-content/uploads/2016/02/Updated_The_PLA_s_New_Organizational_Structure_-_What_is_Known__Unknown_and_Speculation_Parts_1_and_2.pdf

———, "The PLA's New Organizational Structure: What Is Known, Unknown and Speculation (Part 2)," *China Brief*, February 23, 2016. As of July 17, 2017: https://jamestown.org/wp-content/uploads/2016/02/Updated_The_PLA_s_New_Organizational_Structure_-_What_is_Known__Unknown_and_Speculation_Parts_1_and_2.pdf

An Peng [安鹏], "Strategic Consideration on Strengthening the Air Forces in the Maritime Direction" ["加强海上方向空中力量建设的战略思考"], *China Military Science* [中国军事科学], No. 3, 2015, pp. 82–85.

"Asia Today," China Central Television, July 16, 2016.

Axe, David, "China's Increasingly Good Mock Air Battles Prep Pilots for Real War," *Wired*, February 7, 2013. As of July 17, 2017: http://www.wired.com/2013/02/china-mock-air-war/

Beauchamp-Mustafaga, Nathan, Cristina Garafola, Astrid Cevallos, and Arthur Chan, "China Signals Resolve with Bomber Flights over the South China Sea," War on the Rocks, August 2, 2016. As of July 17, 2017: http://warontherocks.com/2016/08/china-signals-resolve-with-bomber-flights-over-the-south-china-sea/

Blanchard, Ben, and Benjamin Kang Lim, "'Give Them a Bloody Nose': Xi Pressed for Stronger South China Sea Response," Reuters, July 31, 2016. As of July 17, 2017: http://www.reuters.com/article/us-southchinasea-ruling-china-insight-idUSKCN10B10G

Brunnstrom, David, "Japan to Boost South China Sea Role with Training Patrols with U.S.: Minister," Reuters, September 15, 2016. As of July 17, 2017: http://www.reuters.com/article/us-southchinasea-japan-patrols-idUSKCN11L2FE

Buckley, Chris, "Chinese General Visits Disputed Spratly Islands in South China Sea," *New York Times*, April 15, 2016. As of July 17, 2017: http://www.nytimes.com/2016/04/16/world/asia/china-south-china-sea-spratly-islands.html

"Build It and They Will Come," Washington, D.C.: Center for Strategic and International Studies, August 1, 2016. As of July 18, 2017: https://amti.csis.org/build-it-and-they-will-come/

Burke, Matthew M., and Chiyomi Sumida, "China Reportedly Responds to Trump's Taiwan Call by Flying Nuclear-Capable Bomber," *Stars and Stripes*, December 12, 2016. As of July 18, 2017: http://www.stripes.com/news/china-reportedly-responds-to-trump-s-taiwan-call-by-flying-nuclear-capable-bomber-1.443999

Cao Chuanbiao [曹传彪] and Zhang Li [张力], "How Many Things Need to Be Coordinated in the Relationship Between Combat Operations and Force Building: The Practice and Thought of the Air Force of the Western Theater Command in Adapting to the New Leadership and Command System and Pushing Forward Realistic Military Training" ["战与建，有多少关系需理顺: 西部战区空军适应新的领导指挥体制推进实战化训练的实践与思考"], *PLA Daily*, October 31, 2016, p. 7. As of July 18, 2017: http://www.81.cn/jfjbmap/content/2016-10/31/content_160312.htm

Chan, Minnie, "PLA Unveils New Weapons for Air and Sea Combat Following Hague Tribunal Ruling," *South China Morning Post*, July 22, 2016. As of July 18, 2017: http://www.scmp.com/news/china/diplomacy-defence/article/1992973/pla-unveils-new-weapons-air-and-sea-combat-following

Chan, Minnie, and Choi Chi-yuk, "Former Top Officer in China's Air Force 'Under Investigation by Anti-Corruption Officials': Sources," *South China Morning Post*, July 8, 2016. July 18, 2017: http://www.scmp.com/news/china/diplomacy-defence/article/1987131/former-top-officer-chinas-air-force-under-investigation

Chase, Michael S., Kenneth W. Allen, and Benjamin S. Purser III, *Overview of People's Liberation Army Air Force "Elite Pilots,"* Santa Monica, Calif.: RAND Corporation,

RR-1416-AF, 2016. As of July 16, 2017: https://www.rand.org/pubs/research_reports/RR1416.html

Chase, Michael S., and Cristina L. Garafola, "China's Search for a 'Strategic Air Force,'" *Journal of Strategic Studies*, Vol. 39, No. 1, September 14, 2015, pp. 4–28.

Chen Dianhong [陈典宏] and He Weili [何伟理], "Shape a 'Joint Battlefield', Foster the 'Strongest Brain': Army Divisions and Brigades of the PLA Southern Theater Command Conduct Authentic and Complex Online Command Confrontation Drills" ["构'联合战场' 练'最强大脑': 南部战区陆军合成师旅网上指挥对抗演练逼真复杂"], *PLA Daily*, May 9, 2016, p. 4. As of July 18, 2017: http://www.81.cn/jfjbmap/content/2016-05/09/content_143963.htm

Chen Haifeng [陈海峰], "Joint Sea 2015 (II) Military Exercise Starts Today, China and Russia Exercise in Sea of Japan" ["海上联合-2015(II)军演今起举行 中俄将在日本海联演"], China News Online, August 20, 2015. As of July 18, 2017: http://www.chinanews.com/mil/2015/08-20/7477927.shtml

Chen Wei-han, "MND Plays Down China Aircraft Photos," *Taipei Times*, December 19, 2016. As of July 18, 2017: http://www.taipeitimes.com/News/front/archives/2016/12/19/2003661487

"China Launches Intensive Training for Several Types of Fighters: Su-30 Flying Leopard Eye-Catching" ["中国多型战机密集升空训练苏 30 飞豹抢眼"], China News Online, March 4, 2015. As of July 18, 2017: http://military.people.com.cn/n/2015/0304/c172467-26637038.html

"China—Air Force," *Jane's World Air Forces*, December 23, 2016.

"China, Russia Start Joint Navy Drill in South China Sea," Xinhua, September 13, 2016. As of July 18, 2017: http://news.xinhuanet.com/english/2016-09/13/c_135685439.htm

"China's Air Force Conducts Combat Air Patrol in South China Sea," Xinhua, July 18, 2016. As of July 18, 2017: http://news.xinhuanet.com/english/2016-07/18/c_135522288.htm

"China's High Sea Military Drills Are Routine: Spokesperson," Xinhua, December 15, 2016. As of July 18, 2017: http://eng.chinamil.com.cn/view/2016-12/15/content_7408806.htm

"Chinese Air Force Combat Aircraft Conduct 'Combat Air Patrol' over South China Sea" ["中国空军多型主战飞机赴南海战斗巡航"], China Central Television, August 6, 2016. As of

July 18, 2017:
http://tv.cctv.com/2016/08/06/VIDEUJCaFLcMjbYjAVpUx7Fe160806.shtml

"Chinese Air Force H-6Ks Conduct Combat Air Patrol over South China Sea and Scarborough Shoal" ["中国空军轰—国空战斗巡航南海黄岩岛"], China Central Television, July 18, 2016. As of July 18, 2017:
https://www.youtube.com/watch?v=lcW38uOGyKk

"Chinese Military Aircraft Patrol South China Sea," Xinhua, August 6, 2016. As of July 18, 2017:
http://news.xinhuanet.com/english/2016-08/06/c_135570247.htm

"Chinese Military Planes Flying to Sea of Japan Prompt Scramble," *Asahi Shimbun*, January 10, 2017. As of July 18, 2017:
http://www.asahi.com/ajw/articles/AJ201701100021.html

Cliff, Roger, John F. Fei, Jeff Hagen, Elizabeth Hague, Eric Heginbotham, and John Stillion, *Shaking the Heavens and Splitting the Earth: Chinese Air Force Employment Concepts in the 21st Century,* Santa Monica, Calif.: RAND Corporation, MG-915/2-AF, 2011. As of July 16, 2017:
https://www.rand.org/pubs/monographs/MG915z2.html

Cozad, Mark R., "PLA Joint Training and Implications for Future Expeditionary Capabilities: Testimony Presented Before the U.S.-China Economic and Security Review Commission, January 21, 2016," Santa Monica, Calif.: RAND Corporation, CT-451, 2016. As of July 21, 2017:
https://www.rand.org/pubs/testimonies/CT451.html

"Defense Review Week," China Central Television, September 11, 2016.

Dong Bin [懂宾], "Courage and Character Shine on the Drill Grounds: Reporter Observations of Air Force Troops Relying on 'System of Systems' Engagement Exercises to Carry Out Intensive Fighting Spirit Training" ["血性在演兵场张:空军部队依托体系对抗演习靠站战斗精神强训之记者观察"], *Kongjun Bao*, June 10, 2015.

Dong Zhaohui, "PLA Air Force Conducts Combat Air Patrol in South China Sea," Xinhua, July 19, 2016. As of July 18, 2017:
http://english.chinamil.com.cn/news-channels/china-military-news/2016-07/19/content_7161691.htm

———, "China's Air Force Conducts West Pacific Drill, Patrols ADIZ," China Military Online, September 25, 2016. As of July 18, 2017:
http://english.chinamil.com.cn/view/2016-09/25/content_7275705.htm

"Draw a Circle Around Taiwan: China's Air Force Flies into Western Pacific from Two Directions" ["给台湾画个圈 中国空军两个方向前出西太平洋训练"], China.com, November 28, 2016. As of July 18, 2017: http://military.china.com/jctp/11172988/20161128/30054512_all.html

Easton, Ian, "China's Evolving Reconnaissance-Strike Capabilities: Implications for the U.S.-Japan Alliance," Arlington, Va.: Project 2049 Institute, February 2014. As of July 18, 2017: http://www.project2049.net/documents/Chinas_Evolving_Reconnaissance_Strike_Capabilities_Easton.pdf

"Exclusive Revelations: 'God of War' H-6K Patrol of Scarborough Shoal" ["独家揭秘"战神轰炸机"轰-6K 巡航黄岩岛"], China Central Television, November 5, 2016. As of July 18, 2017: http://m.news.cctv.com/2016/11/05/ARTISSpdT2nbklcXaNaDWHM2161105.shtml

"Fan Changlong: Continue Deepening Preparations for Military Struggle in the Maritime Direction, Effectively Carry Out the Historic Missions Assigned by the Party and People" ["范长龙:不断深化细化海上方向军事斗争准备 有效履行党和人民赋予使命任务"], Xinhua, July 19, 2016. As of July 18, 2017: http://news.xinhuanet.com/politics/2016-07/19/c_1119245168.htm

Fan Jun [范军] and An Hongxin [安红欣], "Lanzhou Military Region Air Force Brigade's 'Golden Dart' Aviators Seek the Way to Fight and Win" ["向精学胜战突进:兰空航空兵某旅'金飞镖'飞行员谋打赢之道"], China Air Force Online [中国空军网], January 26, 2016. As of July 18, 2017: http://kj.81.cn/content/2016-01/26/content_6869147.htm

Fan Yishu [范以书], Zeng Ke [曾科], and Li Jianwen [李建文], "The Southern Theater Command Air Force Enhances Maritime Combat Capabilities According to the Requirements of Missions and Tasks by Extending the Sphere of Operations from Near Seas to Distant Seas" ["南部战区空军着眼使命任务提升海上作战能力: 跨越: 从近海到远海"], *PLA Daily*, October 13, 2016, p. 4. As of July 18, 2017: http://www.81.cn/jfjbmap/content/2016-10/13/content_158815.htm

Fisher, Richard, "Xi Jinping Visit Reveals H-6 Bomber Details," *IHS Jane's Defence Weekly*, February 19, 2015.

———, "China Showcases New Weapon Systems at 3 September Parade," *Jane's Defence Weekly*, September 8, 2015.

———, "Chinese Bomber Exercise Affirms Air-Defence Identification Zone, Penetrates Second Island Chain," *IHS Jane's 360*, December 1, 2015.

———, "Chinese H-6K Bomber Flies over Fiery Cross Reef in South China Sea," *IHS Jane's 360*, May 12, 2016.

"Focus Today," China Central Television, November 3, 2015.

"Focus Today," China Central Television, November 30, 2015.

"Focus Today," China Central Television, July 18, 2016.

"Focus Today," China Central Television, July 19, 2016. As of July 18, 2017 http://tv.cctv.com/2016/07/20/VIDEOUtxoKEgci5s3vn7WhJW160720.shtml

"Focus Today," China Central Television, August 7, 2016. As of July 18, 2017: http://tv.cctv.com/2016/08/07/VIDERnQrWIG2A6ANgHbrgWkk160807.shtml

Fravel, M. Taylor, "No, Hu Didn't Call for War," *The Diplomat*, December 10, 2011. As of July 18, 2017:
http://thediplomat.com/2011/12/no-hu-didnt-call-for-war/

Fravel, M. Taylor, "China's New Military Strategy: 'Winning Informationized Local Wars,'" *China Brief*, July 2, 2015. As of July 18, 2017:
https://jamestown.org/program/chinas-new-military-strategy-winning-informationized-local-wars/

"From Phoenix to the World," Phoenix TV, May 21, 2015.

Fuell, Lee, "Broad Trends in Chinese Air Force and Missile Modernization," Testimony Before the U.S.-China Economic and Security Review Commission, January 30, 2014. As of July 18, 2017:
http://www.uscc.gov/sites/default/files/Lee%20Fuell_Testimony1.30.14.pdf

Gao Feng [高峰] and Zuo Dengyun [左登云], "Exert Force with Precision, Speed Up the Building of a Maritime Combat Operations System" ["精准发力, 加快海上作战体系建设"], *PLA Daily*, July 31, 2016, p. 7. As of July 18, 2017:
http://www.81.cn/jfjbmap/content/2016-07/31/content_152148.htm

"Guangzhou Air Force H-6K Have Capability to Launch Precision Strikes Anywhere in Chinese Territory Under Any Weather Conditions" ["广空轰-6K 部队具备全疆域全天候远程精确打击能力"], People's Daily Online, October 13, 2015. July 18, 2017:
http://military.people.com.cn/n/2015/1013/c1011-27690474.html

Guo Renjie, "PLA Air Force's Flying over Miyako Strait for Training: Legitimate and Reasonable," China Military Online, May 22, 2015. As of July 18, 2017:
http://english.chinamil.com.cn/news-channels/china-military-news/2015-05/22/content_6503404.htm

"H-6 Bomber Flies over Japan's Southwestern Islands and Enters for First Time" ["轰 6 轰炸机首次飞越日本西南诸岛进入太平洋"], *Observer* [观察者], September 8, 2013. As of July 18, 2017:
http://www.guancha.cn/militaryaffairs/2013_09_08_170989.shtml

He Wu [鹤舞], "Wow! The Air Force Did All This in One Year Since Military Reform" ["厉害!军改一年空军干了这么多大事"], China Military Online, November 25, 2016. As of July 18, 2017:
http://www.81.cn/syjdt/2016-11/25/content_7377278.htm

Heng Jing [靖恒], "Chinese Air Force for the First Time Releases Photos of H-6K Patrolling Scarborough Shoal" ["中国空军首次曝光轰-6K 巡航黄岩岛高清照片"], *Observer*, July 15, 2016. As of July 18, 2017:
http://www.guancha.cn/military-affairs/2016_07_15_367643.shtml

Huang, Kristin, "Japan, South Korea Scramble Fighter Jets After Eight Chinese Warplanes Fly over Korea Strait," *South China Morning Post*, January 10, 2017. As of July 18, 2017:
http://www.scmp.com/news/china/diplomacy-defence/article/2060823/japan-and-south-korea-scramble-fighter-jets-8-chinese

Huang Zijuan [黄子娟], "China's Air Force Tankers Possess Ability to Fly Covertly, Can Fly at Low Sea Levels" ["我军空中加油机具隐蔽出航能力 能海上低空突防"], People's Daily Online, May 18, 2015. As of July 18, 2017:
http://military.people.com.cn/n/2015/0518/c1011-27017116.html

Huang Zijuan [黄子娟], "Chinese Air Force's Combat Ability Had Seven Breakthroughs: Aimed at Winning Future Maritime Warfare" ["中国空军近年七大突破:远海作战能力大幅提升"], People's Daily Online, November 11, 2015. As of July 18, 2017:
http://military.people.com.cn/n/2015/1111/c1011-27800444-2.html

Information Office of the State Council, *China's Military Strategy*, May 26, 2015. As of July 18, 2017:
http://www.chinadaily.com.cn/china/2015-05/26/content_20820628.htm

"J-11 Escort H-6K Fly over Mischief Reef" ["歼-11 护航轰-6K 飞越美济礁 镜头长时间对准中业岛"], Phoenix TV, July 20, 2016.

Japanese Ministry of Defense, "2014 Defense of Japan," 2014. As of July 18, 2017:
http://www.mod.go.jp/e/publ/w_paper/pdf/2014/DOJ2014_1-1-3_web_1031.pdf

———, "On the Flight of Chinese Aircraft in the East China Sea" ["中国機の東シナ海における飛行について"], May 21, 2015. As of July 18, 2017:
http://www.mod.go.jp/js/Press/press2015/press_pdf/p20150521_02.pdf

———, “On the Flight of Chinese Aircraft in the East China Sea” [“中国機の東シナ海における飛行について”], November 27, 2015. As of July 18, 2017: http://www.mod.go.jp/js/Press/press2015/press_pdf/p20151127_02.pdf

———, “2016 Defense of Japan,” August 2, 2016. As of July 18, 2017: http://www.mod.go.jp/j/publication/wp/wp2016/pdf/28010203.pdf

———, “On the Flight of Chinese Aircraft in the East China Sea” [“中国機の東シナ海における飛行について”], September 25, 2016. As of July 18, 2017: http://www.mod.go.jp/js/Press/press2016/press_pdf/p20160925_01.pdf

———, “On the Flight of Chinese Aircraft in the East China Sea” [“中国機の東シナ海における飛行について”], November 28, 2016. As of July 18, 2017: http://www.mod.go.jp/js/Press/press2016/press_pdf/p20161125_02.pdf

———, “On the Flight of Chinese Aircraft in the East China Sea” [“中国機の東シナ海における飛行について”], December 10, 2016. As of July 18, 2017: http://www.mod.go.jp/js/Press/press2016/press_pdf/p20161210_01.pdf

———, “On the Flight of Chinese Aircraft in the East China Sea and Japan Sea” [“中国機の東シナ海及び日本海における飛行について”], January 9, 2017. As of July 18, 2017: http://www.mod.go.jp/js/Press/press2017/press_pdf/p20170109_01.pdf

———, “Emergency Responses in the Third Quarter of FY2016” [“平成２８年度３四半期までの緊急発進実施状況について”], January 20, 2017. As of July 18, 2017: http://www.mod.go.jp/js/Press/press2017/press_pdf/p20170120_02.pdf

Johnson, Jesse, “Chinese State Media Blasts Japan over South China Sea ‘Patrols,’ but Experts See No Change in Policy,” *Japan Times*, September 18, 2016. As of July 18, 2017: http://www.japantimes.co.jp/news/2016/09/18/national/chinese-state-media-blasts-japan-south-china-sea-patrols-experts-see-no-change-policy/

———, “Japan Scrambles Jets as China Warplanes Fly Through Okinawa Strait,” *Japan Times*, September 26, 2016. As of July 18, 2017: http://www.japantimes.co.jp/news/2016/09/26/national/politics-diplomacy/fighter-scrambled-near-okinawa-china-air-force-planes-pass-miyako-strait/#.WPpP5lPyuV4

Ka Weixing [卞卫星], “Tian Xiusi Observes Blue Sky Aerial Competition and Awards” [“田修思观看空装篮球决赛并颁奖”], *Kongjun Bao*, June 16, 2015, p. 1.

Kowalski, James M., “Air Force Global Strike Command,” U.S. Air Force, May 7, 2013. As of July 18, 2017: http://fas.org/programs/ssp/nukes/nuclearweapons/AFGSC-CommandBrief-May2013.pdf

Kui Jun [隗俊], "English Media: PLA Fighters Flew Around Taiwan, Taiwan Says Closely Monitored" ["英媒:解放军战机环绕台湾飞行 台方称严密监控"], *References News* [参考消息], November 28, 2016. As of July 18, 2017: http://news.sina.com.cn/c/nd/2016-11-28/doc-ifxyawmp0410955.shtml

LaGrone, Sam, "China, Russia Land 400 Marines in First Joint Pacific Amphibious Exercise," *USNI*, August 26, 2015. As of July 18, 2017: https://news.usni.org/2015/08/26/china-russia-land-400-marines-in-first-joint-pacific-amphibious-exercise

Li Dapeng [李大鹏], "What Breakthroughs Does Joint Sea-2015 China-Russia Exercise Reveal?" ["海上联合-2015"中俄联演实现了哪些新突破"], China Youth Daily Online [中国青年网], October 9, 2015. As of July 19, 2017: http://zqb.cyol.com/html/2015- 10/09/nw.D110000zgqnb_20151009_2-09.htm

Li Jianwen [李建文], "The Trends of Military Training in the New Year" ["新年度军事训练趋势扫描"], *PLA Daily*, February 1, 2016. As of July 19, 2017: http://www.81.cn/jfjbmap/content/2016-02/01/content_779.htm

Li Jianwen [李建文] and Lu Hui [卢辉], "Aviation Division of the Nanjing Military Region Air Force Uses the Military Parade Spirit to Arouse the Personnel's Combat Readiness Enthusiasm, Painstakingly Practices Skills in Flight Training, Raises Actual Combat Capability" ["苦练精飞提高实战能力"], *PLA Daily*, September 19, 2015. As of July 19, 2017: http://kj.81.cn/content/2015-09/21/content_6690176.htm

Li Jianwen [李建文] and Yu Hongwei [余泓纬], "Multiple Types of Chinese Air Force Aircraft Fly into Western Pacific to Test Distant Sea Combat Capability" ["空军多型战机赴西太平洋检验远海实战能力"], China Military Online, September 13, 2016. As of July 19, 2017: http://www.81.cn/jfjbmap/content/2016-09/13/content_156578.htm

Li Kaiqiang [李开强], "Chinese Air Force Spokesperson: PLA Will Continue Patrols in East China Sea ADIZ" ["空军发言人:解放军将继续在东海防空识别区警巡"], *PLA Daily*, November 5, 2016. July 19, 2017: http://www.81.cn/jwgz/2016-11/05/content_7344224.htm

Li Kaiqiang [李开强] and Wang Zhijia [王志佳], "Cadets Fly Bombers for First Time" ["空军飞行学员首飞喷气式轰炸机"], China Military Online, April 12, 2016. As of July 19, 2017: http://kj.81.cn/content/2016- 04/12/content_7002407.htm

Li Kaiqiang [李开强] and Wu Yi [伍佚], "'Golden Helmet' Kills 'Combat Elite'" ["'金头盔'PK'空战精英':海空军航空兵首次异型机自由空战对抗演练侧记"], *PLA Daily*, October 12, 2015. As of July 19, 2017: http://www.81.cn/jfjbmap/content/2015-10/12/content_125786.htm

Li Kaiqiang [李开强] and Zhang Hongwei [张宏伟], "Air Force Organized First Live Fire Training at Sea for Flight School" ["空军首次组织飞行院校海上实弹打靶"], *PLA Daily*, May 20, 2015. As of July 19, 2017: http://www.mod.gov.cn/academy/2015-05/20/content_4585772.htm

Li Qing [李清], "Chinese Air Force Conducts Golden Dart Ground Attack Exercise for the First Time" ["中国空军首次举行对地攻击"金飞镖"大赛"], China Military Online, October 7, 2014. As of July 19, 2017: http://news.xinhuanet.com/photo/2014-10/07/c_127069612.htm

Li Weisheng [李为胜] and Hao Maojin [郝茂金], "Female Flying Warriors Train over the Territorial Sea: Air Force Aviation Regiment of the Northern Theater Command Carries Out a Maritime Penetration and Assault Exercise" ["飞天女将试剑海疆: 北部战区空军航空兵某团执行海上突防突击演练侧记"], *Kongjun Bao*, May 27, 2016, p. 1.

Liang Fang [梁芳], Post on Sina Weibo [新浪微博], December 12, 2016. As of July 19, 2017: http://m.weibo.cn/status/4051850447972688

Liao Qirong [廖启荣], "Build the Elite Force Along the Course Specified by the Rules: Investigation and Consideration on the Practice of an Aviation Division of the Nanjing Military Region Air Force in Raising the Training Rules Execution Power in the Course of Strictly Conducting Training in Accordance with the Outline" ["雄帅之路依法而行:南空航空兵某师按纲施训从严治训提升训练法规执行力的调查与思考"], *Kongjun Bao*, February 10, 2015.

Lin Yuan [林 源], Yu Hongchun [余红春], Yan Guoyou [闫国有], Chi Yuguang [迟玉光], Bai Zhongbao [白仲宝], and Mu Xingguang [牟星光], "Bombers Take Action" ["轰炸机出动"], *China Air Force* [中国空军], August 19, 2015. As of July 19, 2017: http://kj.81.cn/content/2015-08/19/content_6639556.htm

Liu Feng'an [刘逢安], "Space Net Won't Leak" ["天网恢恢疏而不漏"], *Modern Military* [现代军事], November 2013. As of July 19, 2017: http://www.qikan.com.cn/article/xdjj20131102.html

Liu Guangzhi [刘广智], Tian Xuhui [田旭辉], and Liu Daquan [刘大全], "China's Air Force Mobilizes Three Types and Five Planes to Participate in China-Russia Maritime Exercise"

["中国空军出动 3 型 5 架战机参加中俄海空联合军演"], China Military Online, August 24, 2015. As of July 19, 2017: http://kj.81.cn/content/2015-08/24/content_6646892.htm

Lo, Kinlin, "'It Was a Provocative Act': PLA Pilot Describes Encounter with Foreign Military Jet," *South China Morning Post*, February 13, 2017. As of July 19, 2017: http://www.scmp.com/news/china/diplomacy-defence/article/2070497/it-was-provocative-act-pla-pilot-describes-encounter

Ma Dezu [马德祖], Yang Chunyuan [杨春源], and Zhao Feipeng [赵飞鹏], "Practice to Command a War: Report on the Joint Exercises Conducted by the PLA Air Force Command Academy's New Graduates" ["练习指挥一场战争: 空军指挥学院毕业学员联合作业演习实录"], China Youth Daily Online, January 18, 2008. As of July 19, 2017: http://news.ifeng.com/mil/2/200801/0118_340_369900.shtml

Ma Jun [马俊], "Dissimilar Main Force Combat Aircraft Combat Patrols of South China Sea: Chinese Air Force Shows Its Strength" ["多型主力战机战斗巡航南海!中国空军展示实力"], *Global Times*, August 8, 2016. As of July 19, 2017: http://mil.huanqiu.com/observation/2016-08/9277166.html

Ma Junqing [马俊卿], "Several Types of Air Force and Navy Fighters Participate in 'Sharp Sword-2015' Military Exercises" ["我空军海军多型战机参加'利剑一2015'军实兵演习"], Xinhua, November 3, 2015. As of July 19, 2017: http://news.xinhuanet.com/photo/2015-11/03/c_128389273_19.htm

Ma Xiaotian [马晓天], "Strive to Enhance the Air Force's Capability of Fighting and Winning" ["努力提高空军部队能打仗打胜仗能力"], *PLA Daily* [解放军报], April 2, 2014, p. 6. As of July 19, 2017: http://www.81.cn/2014-content_18620/2014-04/02/content_6153511.htm

Mangosing, Frances, "PH Can't Confirm Chinese Bomber Flight over Scarborough," *Inquirer*, July 19, 2016. As of July 19, 2017: https://globalnation.inquirer.net/141492/ph-no-info-on-chinese-bomber-flight-over-scarborough

McCauley, Kevin, "System of Systems Operational Capability: Key Supporting Concepts for Future Joint Operations," *China Brief*, October 5, 2012. As of July 19, 2017: https://jamestown.org/program/system-of-systems-operational-capability-key-supporting-concepts-for-future-joint-operations/

———, *PLA System of Systems Operations: Enabling Joint Operations,* Washington, D.C.: Jamestown Foundation, 2016.

Military Affairs Decoded [军情解码], "The Theory Behind Russian Bombers' Strategic Cruising" ["俄罗斯轰炸机战略巡航幕后玄机"], February 8, 2013. As of July 19, 2017: http://tv.cntv.cn/video/C30411/a41b368832133ddd528fd8ab0a4e5c44

"Military Report," China Central Television, November 2, 2010.

"Military Report," China Central Television, August 1, 2014.

"Military Report," China Central Television, November 11, 2014.

Mizokami, Kyle, "China's Air Forces Are Venturing Further into the Pacific Than Ever Before," *Popular Mechanics*, September 16, 2016. As of July 19, 2017: http://www.popularmechanics.com/military/weapons/a22861/chinas-air-forces-announce-regular-patrols-into-the-west-pacific/

———, "China's Flyover Message to Trump Wasn't Quite What It Seemed," *Popular Mechanics*, December 12, 2016. As of July 19, 2017: http://www.popularmechanics.com/military/weapons/a24268/china-bombers-south-china-sea-trump/

"MND Appeals for Calm After Reports of Chinese Aircraft Circling Taiwan," *China Post*, December 19, 2016. As of July 19, 2017: http://www.chinapost.com.tw/taiwan/national/national-news/2016/12/19/487089/MND-appeals.htm

Mou Xingguang [牟星光] and Chi Yuguang [迟玉光], "An Aviation Division of the Eastern Theater Air Force Breaks Free of Constrained Thinking, Sets Out on a New Path of Joint Training" ["训练需求表"向实战深度融合: 东部战区空军航空兵某师打破思维局限迈开联合训练新路"], *Kongjun Bao*, April 8, 2016, p. 1.

"Multiple Types of Chinese Air Force Aircraft Conduct First Patrol of East China Sea ADIZ" ["中国空军多架战机首巡东海防空识别区"], *Beijing Morning Post* [北京晨报], November 24, 2013. As of July 19, 2017: http://news.qq.com/a/20131124/001637.htm

"Multiple Types of Chinese Air Force Aircraft Coordinate Defense Penetration and Assault: H-6 and Su-30 [Conduct] Anti-Carrier [Operations]" ["中国空军多机种协同突击突防: 轰 6 苏 30 反航母"], Sina, October 30, 2016. As of July 19, 2017: http://slide.mil.news.sina.com.cn/k/slide_8_193_46266.html#p=1.

"Multiple Types of Chinese Air Force Aircraft Fly Through Bashi Channel and Miyako Strait at Same Time" ["中国空军多型战机同时飞越巴士海峡和宫古海峡"], *Observer*, November 27, 2016. As of July 19, 2017: http://www.guancha.cn/military-affairs/2016_11_27_381950.shtml

"My Story with the God of War," China Central Television, May 6, 2016. As of July 19, 2017: https://www.youtube.com/watch?v=Vty-OQJ5J5g

National Air and Space Intelligence Center, *People's Liberation Army Air Force 2010*, Dayton, Ohio, 2010. As of July 19, 2017: http://www.au.af.mil/au/awc/awcgate/nasic/pla_af_2010.pdf

National Archive CIA Records Search Tool, *China: Military Options Against Vietnam*, March 1, 1984, CIA-RDP84S00928R0003000050006-0, Declassified, p. 10.

National Institute for Defense Studies, *NIDS China Security Report 2016: The Expanding Scope of PLA Activities and the PLA Strategy*, March 2016. As of July 19, 2017: http://www.nids.mod.go.jp/publication/chinareport/pdf/china_report_EN_web_2016_A01.pdf

Office of the Secretary of Defense, *Annual Report to Congress: Military and Security Developments Involving the People's Republic of China 2010*, Arlington, Va.: U.S. Department of Defense, 2010. As of July 19, 2017: https://www.defense.gov/Portals/1/Documents/pubs/2010_CMPR_Final.pdf

———, *Annual Report to Congress: Military and Security Developments Involving the People's Republic of China 2016,* Arlington, Va.: U.S. Department of Defense, April 26, 2016.

"One on One: Liu Rui: The 'God of War' Story" ["[面对面]刘锐：'战神'故事"], China Central Television, February 12, 2017.

Pan, Jason, "Chinese Aircraft Fly Around Taiwan," *Taipei Times*, December 11, 2016. As of July 19, 2017: http://www.taipeitimes.com/News/front/archives/2016/12/11/2003660975

Perlez, Jane, "Tribunal Rejects Beijing's Claims in South China Sea," *New York Times*, July 12, 2016. As of July 19, 2017: https://www.nytimes.com/2016/07/13/world/asia/south-china-sea-hague-ruling-philippines.html

"PLA Air Force Releases Photos of Seemingly H-6K Next to Taiwan Mountain" ["解放军空军发布疑似轰-6K 与台湾中央山脉高峰合影"], *Observer*, December 17, 2016. As of July 19, 2017: http://www.guancha.cn/military-affairs/2016_12_17_384771.shtml

"PLA JL-8 Trainer Aircraft Conducts Maritime Training for First Time" ["我军 JL-8 教练机首次进行海上训练 低空掠过客轮"], China Military Online, September 19, 2014. As of July 19, 2017: http://mil.news.sina.com.cn/2014-09-19/0922801951.html

PLA Military Terms [军语], Beijing, China: Military Science Publishing House [军事科学出版社], December 2011.

"PLA's First Symposium on Military Flight Training Concluded," China Military Online, October 16, 2014. As of July 19, 2017: http://eng.chinamil.com.cn/news-channels/china-military-news/2014-10/16/content_6182824.htm

PLAAF Weibo account, July 15, 2016. As of July 19, 2017: http://www.weibo.cn/single?id=3997501138730865

Pollpeter, Kevin L., Michael S. Chase, and Eric Heginbotham, *The Creation of the PLA Strategic Support Force and Its Implications for Chinese Military Space Operations*, Santa Monica, Calif.: RAND Corporation, RR-2058-AF, 2017. As of November 2017: https://www.rand.org/pubs/research_reports/RR2058.html

Qi Bingxin [齐冰昕], "China's Air Force Conducts Distant Sea Training Through Miyako Strait for First Time" ["中国空军首次飞越宫古海峡开展远海训练"], China Air Force Online, May 22, 2015. As of July 19, 2017: http://www.81.cn/kj/2015-05/22/content_6502556.htm

Qiu Yue [邱越], "Chinese Air Force Distant Sea Training Must Be Regularized, Don't Rule Out Possibility of Foreign Military Surveillance" ["中国空军远海训练必将常态化 不排除遭外军监视可能"], People's Daily Online, March 31, 2015. As of July 19, 2017: http://military.people.com.cn/n/2015/0331/c1011-26775049.html

———, "Expert: Joint Navy and Air Force Exercise Boosts Strength to Seize Air Supremacy and Gives Play to the Navy's Ability to Attack; Chinese Air Force's H-6K Bombers Take Part in 'Sharp Sword-2015 [lijian-2015]' Real-Troop Exercise" ["专家:海空军联演助力制空权争夺 发挥海军打击能力"], People's Daily Online, November 4, 2015. As of July 19, 2017: http://military.people.com.cn/n/2015/1104/c1011-27774164.html

———, "Expert: Chinese Air Force Backbone Can Strike Entire South China Sea, Future Patrols Will Visit More Islands and Reefs" ["专家:中国空军长臂可辐射整个南海 未来将巡航更多岛礁"], People's Daily Online, July 18, 2016. As of July 19, 2017: http://military.people.com.cn/n1/2016/0718/c1011-28563426.html

———, "Air Force Combat Air Patrols in the South China Sea to Be Regularized, Expert: China Has the Capability to Safeguard Sovereignty" ["空军南海战斗巡航将常态化 专家:中国有能力维护主权"], People's Daily Online, July 19, 2016. As of July 19, 2017: http://military.people.com.cn/n1/2016/0719/c1011-28566825.html

———, "Expert: Chinese Air Force Flying into the Distant Seas Forms a System, Initiative Will Expand" ["专家:中国空军赴远海战机形成体系 主动权将扩大"], People's Daily Online, September 28, 2016. As of July 19, 2017: http://military.people.com.cn/n1/2016/0928/c1011-28745109.html

———, "Expert: Air Force Already Has Ability to Respond to Operational Needs in East and South China Seas at Same Time" ["专家:空军已有能力同时应对东海南海作战需求"], People's Daily Online, November 29, 2016. As of July 19, 2017: http://military.people.com.cn/n1/2016/1129/c1011-28904590.html

Qiu Yue [邱越] and Xiao Hong [肖红], "Several Types of Chinese Air Force Fighters Passed 1,000 km Beyond the First Island Chain to Conduct Distant Sea Training" ["中国空军多型战机飞出第一岛链 1000 余公里展开远海训练"], People's Daily Online, August 14, 2015. As of July 19, 2017: http://military.people.com.cn/n/2015/0814/c1011-27463775.html

Qiu Yue [邱越] and Zhao Yanhong [赵艳红], "'Blue Shield-2014' Exercise: An Air Force Missile Brigade Sharpens Its Sword in the Gobi Desert" ["'蓝盾—2014'演习: 空军某地空导弹旅戈壁大漠引弓砺剑"], China Military Online, August 14, 2014. As of July 19, 2017: http://military.people.com.cn/n/2014/0814/c1011-25466021.html

Ren Xu [任旭] and Xu Yeqing [徐叶青], "China-Russia Militaries Successfully Conduct Joint Landing Operation in Exercise" ["中俄两军联合立体登陆演习成功举行"], China Navy Online [中国海军网], August 26, 2015. As of July 19, 2017: http://jz.chinamil.com.cn/n2014/tp/content_6648617.htm

Sanger, David E., and Rick Gladestone, "New Photos Cast Doubt on China's Vow Not to Militarize Disputed Islands," *New York Times*, August 8, 2016. As of July 19, 2017: http://www.nytimes.com/2016/08/09/world/asia/china-spratly-islands-south-china-sea.html

Shou Xiaosong [寿晓松], ed., *The Science of Military Strategy* [战略学], Beijing, China: Military Science Press [军事科学出版社], 2013.

Slavin, Erik, "Does China Really Have a Nuclear-Capable Bomber?" *Stars and Stripes*, December 13, 2016. As of July 19, 2017: http://www.stripes.com/news/does-china-really-have-a-nuclear-capable-bomber-1.444071

Stashwick, Steven, "US Media Over-Hypes Chinese Bomber 'Warning' in South China Sea," *The Diplomat*, December 15, 2016. As of July 19, 2017: http://thediplomat.com/2016/12/us-media-over-hypes-chinese-bomber-warning-in-south-china-sea/

Sun Maoqing [孙茂庆], Xu Zhuangzhi [徐壮志], and Li Xuanliang [李宣良], "Vast Desert Sky Quakes with Thunder: Air Force Test and Training Base Forges Modern Air Force" ["大漠长天听惊雷:空军某试验训练基地淬火现代空军纪实"], Xinhua, December 1, 2008. As of July 19, 2017:
http://news.china.com/zh_cn/news100/11038989/20081201/15214028.html

Sun Xiaochen and Wang Qingyun, "Air Force in Pacific Drill to Test Far-Offshore Strength," *China Daily*, September 26, 2016. As of July 19, 2017:
http://www.chinadaily.com.cn/china/2016-09/26/content_26892057.htm

Sweeney, Pete, "Air Forces of China, Pakistan Launch Joint Training Exercise," Reuters, April 9, 2016. As of July 19, 2017:
http://www.reuters.com/article/us-china-pakistan-defense-idUSKCN0X608B

Taiwan Ministry of National Defense, "R.O.C. Military Press Statement Regarding Close Surveillance of People's Liberation Army Far-Offshore Training," December 10, 2016. As of July 19, 2017:
http://www.mnd.gov.tw/English/Publish.aspx?p=73664&Title=More%20News&style=More%20News

Tan Jie [谭洁], "Nanjing Air Force Relies on Large-Scale Exercise to Accelerate Training Transformation: Air Division Completely Enters Ranks of 1st Class Unit in Military Training" ["南空依托大型演练加速推进训练转变:所属航空兵师全部跨入军事训练一级单位行列"], *PLA Daily*, April 4, 2008.

Tao Shelan [陶社兰], "The Chinese Air Force Strengthens Actual Combat-Oriented Air Fights Competition, and One Batch of 'Golden Helmet' Pilots Emerges" ["中国空军加强实战化空战比武 一批"金头盔"飞行员脱颖而出"], Xinhua, September 8, 2015. As of July 19, 2017:
http://www.chinanews.com/mil/2015/09-08/7511911.shtml

Tate, Andrew, "PLAAF Deploys 40 Aircraft to Western Pacific in Show of Strength," *IHS Jane's 360*, September 27, 2016. As of July 19, 2017:
http://www.janes.com/article/64114/plaaf-deploys-40-aircraft-to-western-pacific-in-show-of-strength

Tian Xiusi [田修思], "Strive to Quicken the Building of a Strong People's Air Force with Integration of Air and Space and with Both Offensive and Defensive Capabilities Around the Fulfillment of the Party's Military Strengthening Goal Under the New Situation" ["紧紧围绕实现党在新形势下的强军目标为加快建设空天一体攻防兼备的强大人民空军而奋斗"], *Kongjun Bao*, June 24, 2014, pp. 1–2.

Tomlinson, Lucas, "China Flies Nuclear-Capable Bomber in South China Sea After Trump Taiwan Call, US Officials Say," Fox News, December 9, 2016. As of July 19, 2017: http://www.foxnews.com/world/2016/12/09/china-flies-nuclear-capable-bomber-in-south-china-sea-after-trump-taiwan-call-us-officials-say.html

Tomlinson, Lucas, and Jennifer Griffin, "China Flew Nuclear-Capable Bombers Around Taiwan Before Trump Call with Taiwanese President," Fox News, December 5, 2016. As of July 19, 2017: http://www.foxnews.com/world/2016/12/05/china-flew-nuclear-capable-bombers-over-taiwan-before-trump-call-with-taiwanese-president.html

Wang Baofei [汪保飞] and Yang Qingyue [杨清悦], "Guard Against 'Loss of Acclimatization' of Combat Mentality" ["谨防作战思维'水土不服'"], *PLA Daily*, December 6, 2015. As of July 19, 2017: http://military.people.com.cn/n/2015/1206/c1011-27893647.html

Wang Dongming [王东明], Zhang Yuqing [张玉清], and Wang Nannan [王楠楠], "What All Did the Chinese Air Force Do This Year: Flew Through the First Island Chain Four Times for Training in the Western Pacific" ["中国空军今年都干了啥:4 次飞出第一岛链演兵西太"], Xinhua, December 29, 2015. As of July 19, 2017: http://inews.ifeng.com/yidian/46873326/news.shtml

Wang Shibin, "Xi Jinping Visits Officers and Men of PLA Air Base," China Military Online, February 5, 2013. As of July 19, 2017: http://en.people.cn/90786/8121701.html

Wang Tonghua [王通化] and Li Kaiqiang [李开强], "For the Country, a Sharp Sword" ["为国仗剑'锐一代'"], *PLA Daily*, December 29, 2016. As of July 19, 2017: http://www.81.cn/jfjbmap/content/2016-12/29/content_165462.htm

Wei Bing [魏兵] and Cao Chuanbiao [曹传彪], "Aviation Division of the Air Force Applies a Task Planning System to Realistic Combat Training According to Actual Missions, Precisely Linking Air Battle Training to the Actual Battlefield" ["空军航空兵某师运用任务规划系统推进实战化训练"], *PLA Daily*, February 18, 2016, p. 1. As of July 19, 2017: http://www.81.cn/jfjbmap/content/2016-02/18/content_1788.htm

Wei Yunfeng [魏云峰], "Taiwan Media Cry Out over PLA Fighters Training Around Taiwan, Taiwan Military Says Under Control" ["台媒惊呼解放军战机绕台演训 台军方称尽在掌握"], *Global Times*, November 28, 2016. As of July 19, 2017: http://mil.huanqiu.com/observation/2016-11/9738328.html

Wilson, Jordan, "China's Expanding Ability to Conduct Conventional Missile Strikes on Guam," U.S.-China Economic and Security Review Commission, May 10, 2016.

Wong, Catherine, "PLA Air Force Holds Fresh Patrols in South China Sea as It Seeks to 'Normalise' Stepped-Up Surveillance," *South China Morning Post*, August 7, 2016. As of July 19, 2017: http://www.scmp.com/news/china/diplomacy-defence/article/2000174/pla-air-force-holds-fresh-patrols-south-china-sea-it

Wood, Peter, "China-Japan Relations Worsen as Chinese Naval Aviation Flies Through 'First Island Chain,'" *China Brief*, November 7, 2013. As of July 19, 2017: https://jamestown.org/program/china-japan-relations-worsen-as-chinese-naval-aviation-flies-through-first-island-chain/

———, "Snapshot: China's Southern Theater Command," *China Brief*, July 22, 2016. As of July 19, 2017: https://jamestown.org/program/snapshot-chinas-southern-theater-command/

Wroe, David, "Chinese Bomber Planes from South China Sea and Future Missiles Could Threaten Australia," *Sydney Morning Herald*, September 3, 2016. As of July 19, 2017: http://www.smh.com.au/federal-politics/political-news/chinese-bomber-planes-from-south-china-sea-and-future-missiles-could-threaten-australia-20160902-gr75yf.html

"XAC H-6," *IHS Jane's 360*, undated.

"Xi Jinping Inspects Air Force Organ" ["习近平在空军机关调研"], Xinhua, April 14, 2014. As of July 20, 2017: http://news.xinhuanet.com/photo/2014-04/14/c_126389555.htm

Xing Hongbo [邢洪波], "The PLA Air Force's Distant Sea Training Is Legal, Legitimate, Justifiable" ["中国空军飞越宫古海峡赴远海训练合法合理合情"], People's Daily Online, May 22, 2015. As of July 20, 2017: http://military.people.com.cn/n/2015/0521/c1011-27037878.html

Xiong Weiwei [熊伟威], "Investigating and Researching at Air Force Organs, Fan Changlong, Member of the CPC Central Committee Political Bureau and Vice Chairman of the Central Military Commission, Stresses the Serious Implementation of the Series of Important Decisions and Instructions of Chairman Xi and Making Persistent Efforts in Deeply Pushing Forward Actual Combat-Oriented Training: Fang Fenghui, Member of the Central Military Commission and Chief of General Staff, Conducts Research and Study, Ma Xiaotian, Qi Jianguo, Wu Changde, Liu Zheng, Liu Sheng, and Tian Xiusi Also Participate in the Research" ["中共中央政治局委员、中央军委副主席范长龙在空军机关调研时强调:认真

贯彻习主席系列重要决策指示再接再厉深入推进实战化训练"], *Kongjun Bao*, October 27, 2014, p. 1.

Xu Bianchun [徐边春] and Yang Chaoli [杨朝钊], "While Part of the Force Returns from Triumphant Air Battle, the Other Part of the Force Sets Off to Maritime Training" ["这边空战凯歌还 那边海训再出征"], *PLA Daily*, December 17, 2015, p. 9. As of July 20, 2017: http://www.81.cn/jfjbmap/content/2015-12/17/content_132541.htm

Xu Sheng [徐生], "War Eagle Tough Training for Hard Power Signifies Blue Sky" ["战鹰振翅苦练硬功显威蓝天"], China Military Online, January 3, 2016.

Xu Sheng [徐生] and Chi Yuguang [迟玉光], "Aviation Division of the Air Force of the Eastern Theater Command Normalizes Joint Training" ["东部战区空军航空兵某师:联合训练走向常态化"], *PLA Daily*, June 6, 2016, p. 1. As of July 20, 2017: http://www.81.cn/jfjbmap/content/2016-06/06/content_146663.htm

Xu Sheng [徐生] and Yu Hongwei [余泓纬], "A Unit of the Guangzhou Military Region Air Force Forges All-Weather, Whole-Territory Combat Capability with New-Model Strategic Bombers" ["广空某部全天候全疆域锤炼新型战略轰炸机作战能力: 战鹰振翅苦练硬功显威蓝天"], *PLA Daily*, January 3, 2016, p. 1. As of July 20, 2017: http://military.people.com.cn/n1/2016/0103/c172467-28005089.html

Xu Sheng [徐生], Zhao Lingyu [赵凌宇], and Wang Wenbin [王文彬], "Shocking the Skies by Piloting the 'Gods of War'" ["驾'战神'以撼苍穹"], *PLA Daily*, December 8, 2015, p. 5. As of July 20, 2017: http://kj.81.cn/content/2015-12/08/content_6804489.htm

Xu Tongxuan [徐同宣], "Where Does Training Go from Here? Viewing Realistic Battle Training During Informatized Transformation from the Perspective of Air Force Defense Penetration and Assault Competitive Assessments" ["训练向哪里去:从空军突防突击竞赛性考核看信息化转型时期实战化训练"], *Kongjun Bao*, July 16, 2015, p. 3.

———, "PLA Air Force Holds a Meeting in Beijing to Review and Reflect upon Performance in the Maritime Penetration and Assault Contest and Evaluation Events, Stressing the Settlement of Crucial and Difficult Issues in Force Transformation and Building, Calling for More Elaborate Training Guidance" ["空军在京召开海上突防突击竞赛考核论反思会强调:解救转型建设重难点问题提高训练指导精细化水平,马晓天于忠福全程参与交流研讨并讲话"], *Kongjun Bao*, September 17, 2015, p. 1.

———, "It Is Particularly Essential to Step Up Military Training and Combat Readiness in Times of Reform and Tackling Tough Issues: Sidelights on the PLA Air Force's 2016

Military Training Conference" ["越是改革攻坚越要练兵备战: 空军 2016 年度军事训练会议侧记"], *Kongjun Bao*, January 6, 2016, p. 1.

Xue Xiaoling [薛小岭], "Shenyang PLAAF Unit Conducts Flight Training" ["沈空航空兵某旅进行飞行训练"], China Military Online, December 15, 2015. As of July 20, 2017: http://www.81.cn/kj/2015-12/15/content_6816506.htm

Xue Xiaoling [薛小岭] and Chen Yi [陈益], "Combat-Realism Becomes Routine: Shenyang MRAF Aviation Brigade Enhances Training Quality of Over-Land and At-Sea Defense Penetration and Attacks" ["实战化进入常态化: 沈空航空兵某旅提升陆地海上突防突击训练质效"], *Kongjun Bao*, June 9, 2015, p. 2.

Yan Xinyu [严新宇] and Yang Pan [杨盼], "Northern Theater Command Air Force Aviation Division Coordinates with Multiple Types of Military Aircraft in Defense Penetration and Assault Drills" ["北部战区空军航空兵某师: 协同多形战机演习突防突击"], *Kongjun Bao*, June 6, 2016, p. 1.

Yang Qingchun [杨庆春], "When Inspecting a Test-Oriented Training Base, Ma Xiaotian, Member of the Central Military Commission and Commander of the PLA Air Force, Emphasizes Accurately Determining and Deeply Understanding the Base's Functions and Tasks, Scientifically Planning and Comprehensively Consolidating the Foundation for Building and Development" ["准确定位深刻理解基地职能任务科学筹划全面穷实建设发展根"], *Kongjun Bao*, December 23, 2014, p. 1.

———, "While Inspecting a Certain Guangzhou Military Region Air Force Unit, Ma Xiaotian, Member of the Central Military Commission and Commander of the Air Force, Emphasized That Joint Maritime Search and Rescue Will Require Integration of Military and Local Organizations and High Command Efficiency" ["中央军委委员,空军司令员马晓天在广空部队检查调研时强调:海上联合搜救要军地-体多元融合指挥高效"], *Kongjun Bao*, November 10, 2015.

Yang Weike [杨伟科] and Li Jianwen [李建文], "'Golden Dart' Competition Starts Again" ["'金飞镖'争夺战烽烟再起"], *PLA Daily*, July 23, 2016. As of July 20, 2017: http://www.mod.gov.cn/power/2016-07/23/content_4699155.htm

Yang Zhen [杨振] and Li Kaiqiang [李开强], "Chinese Air Force for the First Time Conducts Western Pacific Training" ["中国空军首次赴西太平洋开展远海训练"], *PLA Daily*, March 30, 2015. July 20, 2017: http://jz.chinamil.com.cn/gd/2015-03/31/content_6421328.htm

Yao Jianing, "PLA Air Force Conducts First Training in West Pacific," China Military Online, March 30, 2015. As of July 20, 2017:

http://english.chinamil.com.cn/news-channels/china-military-news/2015-03/30/content_6420527.htm

———, "H-6K Bomber, Ace of PLA Air Force," People's Daily Online, June 6, 2016. July 20, 2017: http://english.chinamil.com.cn/news-channels/china-military-news/2016-06/06//content_7087946.htm

———, "Expert: PLA Air Force Can Cover Whole South China Sea," China Military Online, July 19, 2016. As of July 20, 2017: http://english.chinamil.com.cn/news-channels/pla-daily-commentary/2016-07/19/content_7163246.htm

Ye Pan [叶攀], "40 Aircraft of Multiple Types Go into Western Pacific, China's Air Force Tests Distant Sea Combat Capability" ["40 多架各型战机前出西太平洋 中国空军检验远海实战能力"], Xinhua, September 25, 2016. As of July 20, 2017: http://www.chinanews.com/mil/2016/09-25/8014191.shtml

Yu Guozhong [于国忠] and Xu Tongxuan [徐同宣], "The PLA Air Force Convenes Its Semi-Annual Video Teleconference on Military Training; Ma Xiaotian, Member of the Central Military Commission and Air Force Commander, and Air Force Political Commissar Tian Xiusi Are Present" ["空军召开半年军事训练电视电话会议:中央军委委员，空军司令员马晓天，空军政治委员天修思出席"], *Kongjun Bao*, June 1, 2014.

Yu Hongwei [余泓纬], "Guangzhou H-6K Possesses Ability to Conduct Long-Distance Precision Strikes Anywhere in Chinese Territory" ["广空轰-6K 部队具备全疆域全天候远程精确打击能力"], China Military Online, September 25, 2015. As of July 20, 2017: http://www.chinanews.com/mil/2015/09-30/7552031.shtml

———, "Guangzhou H-6K Possesses Ability to Conduct Long-Distance Precision Strikes Anywhere in Chinese Territory" ["广空轰-6K 部队具备全疆域全天候远程精确打击能力"], China Military Online, October 12, 2015.

Yu Hongwei [余泓伟] and Ren Bin [任斌], "China's Air Force Goes to Western Pacific for Distant Sea Training" ["中国空军赴西太远海训练 飞出第一岛链上千公里"], China News Online, August 14, 2015. As of July 20, 2017: http://www.chinanews.com/m/mil/2015/08-14/7468270.shtml

Yu Hongwei [余泓纬] and Xu Sheng [徐生], "Guangzhou Air Force H-6K Have Capability to Launch Precision Strikes Anywhere in Chinese Territory Under Any Weather Conditions" ["广空轰-6K 部队具备全疆域全天候远程精确打击能力"], China News Online, September

30, 2015. As of July 20, 2017:
http://www.chinanews.com/m/mil/2015/09-30/7552031.shtml

Yu Jinhu [余金虎], "Ten Series of Military Exercises Have Promoted the Innovation and Upgrading of PLA's Combat Effectiveness" ["十大系列军演推动解放军战斗力创新升级"], China Military Online, March 12, 2016. As of July 20, 2017:
http://www.81.cn/jfjsdjy/2016-03/12/content_6956725.htm

Yue Mingliang [岳明亮], Su Zhibo [苏志波], and Ding Yibo [丁义波], "Practical Training: 'Refueling +;' Account of Eastern Theater Command Air Force Air Brigade Organizing Key-Element Integrated Training with Tactical Backdrop" ["实训: '加油+': 东部战区空军航空兵某旅组织战术背景下要素融合训练侧记"], *Kongjun Bao*, August 4, 2016, p. 2.

Zhang Heng [张恒], "Confrontation Between Different Models of Aircraft Is Carried Out in a 'Back-to-Back' Way: The Joint Training of the Aviation Forces of the Navy and the Air Force Is Expanding in Depth" ["异型机对抗背靠背进行:海空军航空兵联合训练向深度拓展"], *PLA Daily*, September 20, 2015. As of July 20, 2017:
http://www.81.cn/rd/2015-09/20/content_6690174.htm

Zhang Junshe [张军社], "Japan Needs to Get Used to Chinese Air Force's Distant Sea Training" ["日本应习惯中国空军远海训练"], *People's Daily Overseas Edition* [人民日报海外版], April 1, 2015, p. 1. As of July 20, 2017:
http://opinion.people.com.cn/n/2015/0401/c1003-26780623.html

———, "China-Russia Naval Exercise Reflects Strategic Mutual Trust Between the Two Countries" ["中俄海上军演展现两国战略互信"], *People's Daily Overseas Edition* [人民日报海外版], September 13, 2016, p. 1. As of July 20, 2017:
http://opinion.people.com.cn/n1/2016/0913/c1003-28710890.html

Zhang Kejin [张科进] and Chen Guoquan [陈国全], "Sino-Russian 'Joint Sea-2016' Exercise Kick Off Today" ["中俄'海上联合-2016'演习今起举行"], *PLA Daily*, September 12, 2016. As of July 20, 2017:
http://www.81.cn/jmywyl/2016-09/12/content_7254121.htm

Zhang Tao, "Expert: China's Bombers Embrace New Opportunities," China Military Online, February 25, 2015. As of July 20, 2017:
http://english.chinamil.com.cn/news-channels/china-military-news/2015-02/25/content_6366685.htm

———, "Expert: H-6K Demonstrates Long-Range Strike Capability," China Military Online, November 4, 2015. As of July 20, 2017:

http://english.chinamil.com.cn/news-channels/china-military-news/2015-11/04/content_6754675.htm

———, "Expert: China Capable of Defending East China Sea ADIZ," China Military Online, December 2, 2015. As of July 20, 2017: http://english.chinamil.com.cn/news-channels/china-military-news/2015-12/02/content_6796532.htm

———, "PLA Air Force's South China Sea Patrol a 'New Normal,'" China Military Online, August 25, 2016. As of July 20, 2017: http://english.chinamil.com.cn/view/2016-08/25/content_7224861.htm

———, "Chinese and Russian Troops Ready for Joint Sea 2016 Drill," China Military Online, September 13, 2016a. As of July 20, 2017: http://eng.chinamil.com.cn/view/2016-09/13/content_7256371.htm

———, "Chinese Military Aircraft in West Pacific for Combat Simulation Drill," China Military Online, September 13, 2016b. As of July 20, 2017: http://english.chinamil.com.cn/view/2016-09/13/content_7254635.htm

Zhang Yuliang [张玉良], ed., *The Science of Campaigns* [战役学], Beijing, China: National Defense University Press [国防大学出版社], 2006.

Zhang Yuqing [张玉清] and Wu Yi [伍轶], "Chinese Air Force's Planes of Several Models Including H-6K Go to Western Pacific Ocean for Blue Water Training; H-6K, Fighter Planes and Early Warning Planes of Several Models Simultaneously Dispatched to Conduct Policing Patrols in East China Sea Air Defense Identification Zone" ["中国空军轰-6K 等多型飞机赴西太平洋远海训练 同时出动轰-6K 和多型歼击机预警机警巡东海防空识别区"], Xinhua, November 27, 2015.

Zhang Yuqing [张玉清] and Zhang Mimi [张汨汨], "China's Air Force Strengthens Distant Sea Training, Increases Strategic Capabilities" ["中国空军加强远海训练提升战略能力"], Xinhua, December 15, 2016. As of July 20, 2017: http://news.xinhuanet.com/politics/2016-12/15/c_1120123755.htm

Zhang Yuwing [张玉清] and Zhao Wei [赵蔚], "China's Air Force Spokesperson Shen Jinke Introduces East China Sea Patrols" ["空军新闻发言人申进科介绍东海警巡"], Xinhua, August 7, 2014. As of July 20, 2017: http://www.scio.gov.cn/xwfbh/jjxwfyr/wz/Document/1377471/1377471.htm

Zhang Zhaoying [张骄瀛], "Expert: China's Air Force Distant Sea Training Has Major Transformation, Air Combat System Already Formed" ["专家:中国空军远海训练有重大变

化 空战体系已形成”], People’s Daily Online, December 16, 2016. As of July 20, 2017: http://news.xinhuanet.com/mil/2016-12/16/c_129407900.htm

Zhang Zhongkai [张钟凯] and Wu Dengfeng [吴登峰], “Military Expert Explains Highlights of China-Russia ‘Joint Sea-2015 (II)’ Exercise” [“军事专家详解中俄‘海上联合-2015(II)’演习亮点”], Xinhua, August 24, 2015. As of July 20, 2017: http://news.xinhuanet.com/mil/2015-08/24/c_1116357242.htm

Zhao Lei, “PLA Shifts Bombers to Flight-Training Role,” *China Daily*, May 13, 2016. As of July 20, 2017: http://usa.chinadaily.com.cn/china/2016-05/13/content_25249328.htm

———, “Air Force’s Patrols to Continue over East, South China Seas,” *China Daily*, December 16, 2016. As of July 20, 2017: http://www.chinadaily.com.cn/china/2016-12/16/content_27687743.htm

Zhou Jianhua [周建华] and Tian Yu [田宇], “Air Force Division of the Central Theater Command Comprehensively Forges Power Projection Capabilities by Challenging Maritime Training Extremes” [“中部战区空军某师全方位锤炼投送能力: 海上训练摸索挑战极限”], *PLA Daily*, August 17, 2016, p. 12. As of July 20, 2017: http://www.81.cn/jfjbmap/content/2016-08/17/content_153819.htm

Zhou Shiyu [周士瑜], “From ‘Training for Fighting’ to ‘Training Being Fighting’: Sidelights of Realistic Combat Drills Between Opposing Forces Conducted by an Aviation Brigade of the Western Theater Command’s Air Force” [“从‘训为战’到‘训即战’: 西部战区空军航空兵某旅组织实战化对抗训练侧纪”], *Kongjun Bao*, August 3, 2016, p. 3.

Zhu Hui [朱晖], ed., *Strategic Air Force* [战略空军论], Beijing, China: Lantian Publishing House [蓝天出版社], 2009.